Collins

Year 8,

NEW MATHS FRAMEWORKING

Matches the revised KS3 Framework

Kevin Evans, Keith Gordon, Trevor Senior, Brian Speed

William Collins' dream of knowledge for all began with the publication of his first book in 1819. A self-educated mill worker, he not only enriched millions of lives, but also founded a flourishing publishing house. Today, staying true to this spirit, Collins books are packed with inspiration, innovation and practical expertise. They place you at the centre of a world of possibility and give you exactly what you need to explore it.

Collins. Freedom to teach.

Published by Collins
An imprint of HarperCollins*Publishers*
77–85 Fulham Palace Road
Hammersmith
London
W6 8JB

Browse the complete Collins catalogue at
www.collinseducation.com

© HarperCollins*Publishers* Limited 2008

10 9 8 7 6

ISBN 978-0-00-726799-6

Keith Gordon, Kevin Evans, Brian Speed and Trevor Senior assert their moral rights to be identified as the authors of this work.

All rights reserved. No part of this publication may be reproduced, stored in a retrieval system, or transmitted in any form or by any means, electronic, mechanical, photocopying, recording or otherwise, without the prior written permission of the Publisher or a licence permitting restricted copying in the United Kingdom issued by the Copyright Licensing Agency Ltd., 90 Tottenham Court Road, London W1T 4LP.

British Library Cataloguing in Publication Data
A Catalogue record for this publication is available from the British Library.

Commissioned by Melanie Hoffman and Katie Sergeant
Project management by Priya Govindan
Covers management by Laura Deacon
Edited by Brian Ashbury
Proofread by Amanda Dickson
Design and typesetting by Newgen Imaging
Design concept by Jordan Publishing Design
Covers by Oculus Design and Communications
Illustrations by Tony Wilkins and Newgen Imaging
Printed and bound by Printing Express, Hong Kong
Production by Simon Moore

Every effort has been made to trace copyright holders and to obtain their permission for the use of copyright material. The authors and publishers will gladly receive any information enabling them to rectify any error or omission in subsequent editions.

FSC is a non-profit international organisation established to promote the responsible management of the world's forests. Products carrying the FSC label are independently certified to assure consumers that they come from forests that are managed to meet the social, economic and ecological needs of present and future generations.

Find out more about HarperCollins and the environment at
www.harpercollins.co.uk/green

Introduction

Welcome to *New Maths Frameworking*!

New Maths Frameworking Year 8 Practice Book 2 has hundreds of levelled questions to help you practise Maths at Levels 5-6. The questions correspond to topics covered in Year 8 Pupil Book 2 giving you lots of extra practice.

These are the key features:

- **Colour-coded National Curriculum levels** for all the questions show you what level you are working at so you can easily track your progress and see how to get to the next level.

- **Functional Maths** is all about how people use Maths in everyday life. Look out for the Functional Maths icon (FM) which shows you when you are practising your Functional Maths skills.

Contents

CHAPTER 1 Number and Algebra **1** — 1

CHAPTER 2 Geometry and Measures **1** — 6

CHAPTER 3 Statistics **1** — 11

CHAPTER 4 Number **2** — 15

CHAPTER 5 Algebra **2** — 19

CHAPTER 6 Geometry and Measures **2** — 23

CHAPTER 7 Algebra **3** — 29

CHAPTER 8 Number **3** — 33

CHAPTER 9 Geometry and Measures **3** — 38

CHAPTER 10	Algebra 4	42
CHAPTER 11	Statistics 2	46
CHAPTER 12	Number 4	51
CHAPTER 13	Algebra 5	54
CHAPTER 14	Solving Problems	59
CHAPTER 15	Geometry and Measures 4	63
CHAPTER 16	Statistics 3	69

CHAPTER 1 Number and Algebra 1

Practice

1A Multiplying and dividing negative numbers

1 Calculate the following.
  a 5 − 9
  b 2 − 6 + 3
  c −2 + −5
  d 9 − −7
  e −3 − +5 − 4

2
  i Starting at **a**, add each number to the number on its left. Write down your answers.
  ii Again starting at **a**, subtract each number from the number on its left. Write down your answers.

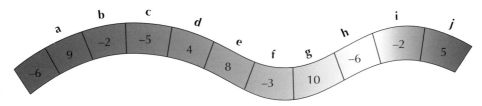

3 Calculate the following.
  a 5 × −3
  b −8 × −5
  c −3 × 10
  d −2 × −2
  e 14 × −3
  f −4 × −2 × 5
  g 2 × −3 × −1
  h −10 × 6 × −10

4 Calculate the following.
  a 16 ÷ −2
  b −6 ÷ −6
  c −9 ÷ 3
  d 12 ÷ −3
  e −100 ÷ 10
  f 20 ÷ −2 ÷ −5
  g −36 ÷ −3 ÷ −4
  h 40 ÷ −5 ÷ −2

5 Copy and complete the following chains of calculations.

a

b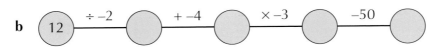

c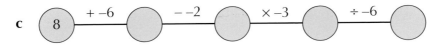

6
  a Write down five calculations involving × that give the answer −6.
  b Write down five calculations involving ÷ that give the answer −5.

7 Calculate the following.

  a $(-8)^2$
  b $4 + -2 \times 7$
  c $-14 \div 2 + 5$
  d $2 \times -6 \div -3$
  e $4 \times (-3 - 5)$
  f $-8 - (7 + 5)$
  g $(-3 - 12) \div -3$

8 Put brackets in to make each of these calculations true.

  a $9 \times -2 + 1 = -9$
  b $-4 + -6 \div -2 = -1$
  c $5 - 7 + 2 - 4 = -8$

Practice — 1B HCF and LCM

1 **a** Write the first 10 multiples of the following numbers.
  i 3 **ii** 6 **iii** 8 **iv** 15

  b Use your answers to find the LCM of the following pairs of numbers.
  i 3 and 8 **ii** 6 and 15 **iii** 6 and 8 **iv** 8 and 15

2 **a** Write out all the factors of the following numbers.
  i 12 **ii** 18 **iii** 20 **iv** 30

  b Use your answers to find the HCF of the following pairs of numbers.
  i 12 and 18 **ii** 12 and 20
  iii 20 and 30 **iv** 18 and 20

3 Find the LCM of the following pairs of numbers.

  a 4 and 8 **b** 6 and 10 **c** 7 and 8 **d** 9 and 12

4 Find the HCF of the following pairs of numbers.

  a 14 and 35 **b** 8 and 20 **c** 12 and 30 **d** 15 and 24

Practice — 1C Powers and roots

1 Find the cubes of these numbers. Show your working.

  a 5 **b** 11 **c** 12

2 Use your calculator to find the following.

  a 24^2 **b** 15^3 **c** 25^3 **d** 6.6^2 **e** 4.2^3 **f** 7.3^3

3 Use your calculator to find the following.

  a 2^6 **b** 4^5 **c** 8^4 **d** 6^6 **e** 3^8

4 Write down the values of the following roots.

  a $\sqrt{16}$ **b** $\sqrt{81}$ **c** $\sqrt[3]{64}$ **d** $\sqrt[3]{343}$

5 Find two values of x that make each equation true.

  a $x^2 = 49$ **b** $x^2 = 100$ **c** $x^2 = 225$ **d** $x^2 = 1.44$

6 **a** Use your calculator to find the following.
  i 0.4^2 **ii** 0.5^2
 b Work out the answer to 0.6^2 in your head. Now check the answer with your calculator.
 c Work out the answer to 0.3^2 in your head. Again check the answer with your calculator.
 d Copy and complete the table below.

Number	0.1	0.2	0.3	0.4	0.5	0.6	0.7	0.8	0.9	1
Square										

7 **a** Choose any number between 0 and 1, for example, 0.4.
 b Calculate increasing powers of your number, for example, 0.4^2, 0.4^3, 0.4^4.
 c What do you notice about the sizes of your answers as the power increases? Explain why this happens.

8 **a** Use your calculator to find the following.
  i 0.5^3 **ii** 0.6^3
 b Copy and complete the table below.

Number	0.1	0.2	0.3	0.4	0.5	0.6	0.7	0.8	0.9	1
Cube										

Practice
1D Prime factors

1 Calculate these products of prime factors.
 a $2 \times 5 \times 5$ **b** $2^3 \times 3$ **c** $2 \times 3 \times 7^2$

2 Use a prime factor tree to write each of the following numbers as a product of its prime factors.
 a 6   **b** 18   **c** 32   **d** 70   **e** 36

3 Use the division method to write each of the following numbers as a product of its prime factors.
 a 14   **b** 45   **c** 96   **d** 130   **e** 200

4 Find the prime factors of all the numbers from 21 to 30.

5 **a** Which numbers in question 4 only have one prime factor?
 b What special name is given to these numbers?
 c What is the next number after 30 with only 1 prime factor?

6 The prime factors of 60 are $2 \times 2 \times 3 \times 5 = 2^2 \times 3 \times 5$.
 a Write down the prime factors of 120 in index form.
 b Write down the prime factors of 180 in index form.
 c Write down the prime factors of 600 in index form.

Practice
1E Sequences 1

1 Follow the instructions in the following flow diagrams to generate sequences.

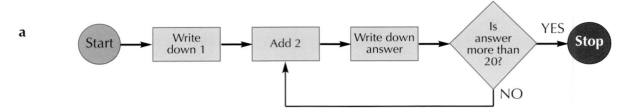

2 a What is the term-to-term rule for each of the following sequences?
  i 5, 8, 11, 14, 17 …   **ii** 2, 10, 50, 250 …
  b Find the next two numbers in each sequence.

3 You are given a start number and a multiplier. Write down the first five terms of the sequence.

  a Start 1, multiplier 20   **b** Start 64, multiplier $\frac{1}{4}$
  c Start −3, multiplier −2

4 Look at the sequences below. The number of dots in each pattern is written below it.

  i Draw the next two patterns of dots in each sequence.
  ii Write down the next four numbers in each sequence.

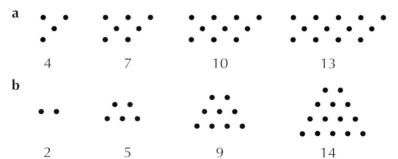

Practice
1F Sequences 2

1 For each of the following arithmetic sequences, write down the first term a and the constant difference d.

  a 6, 8, 10, 12, 14 …   **b** 30, 35, 40, 45 …   **c** 100, 96, 92, 88 …

2 Given the first term a and the constant difference d, write down the first six terms of each sequence.

   a $a = 7, d = 6$ **b** $a = 2, d = 2.5$ **c** $a = 8, d = -5$

3 The following diagram can be used to generate sequences.

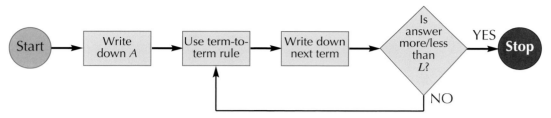

Write down the sequences generated using the following term-to-term rules and values of A and L.

	A	Term-to-term-rule	L
a	2	Treble	100
b	20	Subtract 3	−10
c	1	Add 2, 4, 6, 8, etc.	30
d	1	Multiply by 4, subtract 10	−600
e	500	Divide by 10	0.001

4   **i**   The nth term of each sequence is given below. Write down the first five terms.
    ii   What is the constant difference in each sequence?

   a $3n - 1$ **b** $2n + 5$ **c** $5n - 3$ **d** $10n + 10$

5 Find the nth term of each of the following sequences.

   a   7, 13, 19, 25, 31, …   **b**   5, 8, 11, 14, 17, …
   c   8, 14, 20, 26, 32, …   **d**   1, 4, 7, 10, 13, …
   e   16, 23, 30, 37, 44, …

6 Write down a first term A and a term-to-term-rule that you can use in the flow diagram in question 3 so that:

   a   each term of the sequence is a negative odd number
   b   the first three terms are positive, the remainder are negative
   c   the first three terms are whole numbers, the remainder are decimals.

Practice

1G Solving problems

Carry out the following investigations.

1 Paving slabs 1 metre square are used for borders around L-shaped ponds.

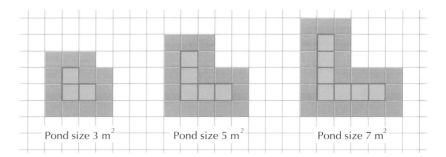

Pond size 3 m² — Pond size 5 m² — Pond size 7 m²

    a How many slabs would fit around a pond of size 9 square metres?
    b Write a rule to show the number of slabs needed to make a border around L-shaped ponds.

2) Write a rule to show the number of matches needed to make the following shapes.

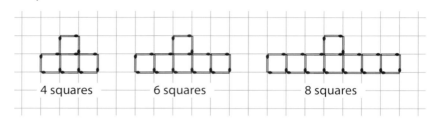

4 squares — 6 squares — 8 squares

3) Martha opened a post office account on Monday. She deposited £8.
On Friday she withdrew half the money in her account.
Every Monday she deposited £4 more than she did the previous Monday.
Every Friday she withdrew half the money in her account.

    a Make a table of the amount of money in her account on the first six Saturdays.
    b Write a rule for the amount of money in her account on a Saturday.

CHAPTER 2 Geometry and Measures 1

Practice

2A Alternate and corresponding angles

1) Copy and complete the following sentences.

    a f and ___ are alternate angles.
    b c and ___ are corresponding angles.
    c ___ and a are corresponding angles.
    d c and e are _____ angles.
    e d and h are _____ angles.

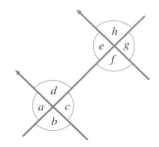

2 **a** Write down three pairs of alternate angles.
b Write down three pairs of corresponding angles.

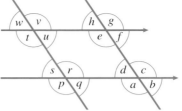

3 Copy each of the following diagrams. Calculate the sizes of all the angles and mark them on your diagram.

a

b

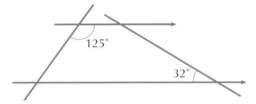

c

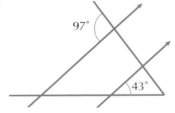

Practice 2B Angles in triangles and quadrilaterals

Find the size of each lettered angle. Find *a* first, then *b* and so on.

1

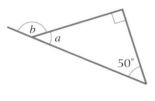

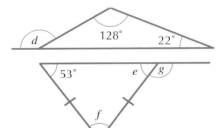

2

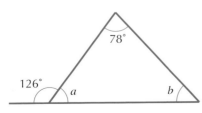

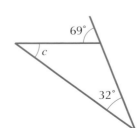

③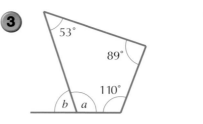

④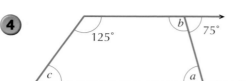

Practice
2C Geometric proof

1. Write a proof to show that opposite angles are equal, that is, $a = c$ and $b = d$. (Hint: Use angles on a straight line.)

   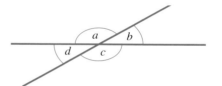

2. ABC is an isosceles triangle. AD bisects angle A. Write out a proof to show that AD is perpendicular to BC, that is, $b = c = 90°$.

   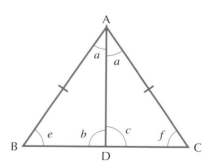

3. Write a proof to show that the opposite angles of a parallelogram are equal.

   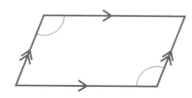 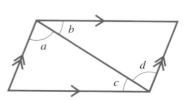

   (Hint: Draw a diagonal, as shown on the right.)

Practice 2D The geometric properties of quadrilaterals

1) Copy the table below. In each column, write the names of all possible quadrilaterals that could fit the description. (Hint: the same quadrilateral could be in more than one column!)

Two pairs of equal angles	Rotational symmetry of order 4	Exactly two lines of symmetry	Exactly two right angles	Exactly four equal sides

2) A quadrilateral has two pairs of equal sides and three obtuse angles. What type of quadrilateral is it?

3) A quadrilateral has three equal angles. What type of quadrilateral could it be?

4) Which quadrilaterals have or could have diagonals that intersect at right angles? Illustrate your answers with drawings.

5) The diagonals of a quadrilateral divide it into four small triangles. Which quadrilaterals contain at least one small isosceles triangle when divided like this? Illustrate your answers with drawings.

Practice 2E Constructions

1) **a** Draw a line AB, 13 cm long. Use a ruler and compasses only to bisect the line. Check the bisection using your ruler.
 b Draw another line AB, 13 cm long. Mark point C on the line 5 cm from C. Construct the perpendicular to AB through point C.

2) Trace the diagram below. Construct a perpendicular from point C to the line AB. Use a ruler and compasses only.

• C

A ———————————————————————— B

3 Use a protractor to draw a 64° angle. Use a ruler and compasses to bisect the angle. Check the angles using your protractor.

4 Repeat question 2 for the angle 134°.

5 a Draw a line XY, 12 cm long.
  b Label the point 5 cm from X as point A.
  c Draw a perpendicular to XY through point A.
  d By measuring the length of the perpendicular, draw a kite with diagonals of length 12 cm and 4 cm.

6 a Place your outstretched hand on a sheet of paper.
Mark the end of your thumb, middle finger and little finger.
Label the points A, B and C respectively.

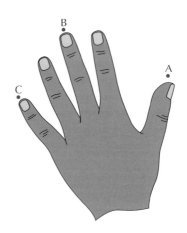

  b Draw the line AC. Bisect it using a ruler and compasses. Label the midpoint D.
  c Compare the width of your closed hand to the length CD.
  d Construct a perpendicular from point B onto AC. Label the intersection E.

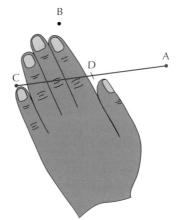

CHAPTER 3 Statistics 1

Practice

3A Probability

1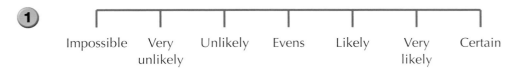

Copy the above scale. Label the scale with each of the following events.

- **a** You pick one card from a pack of 52 cards and it is an ace (there are four aces in a pack).
- **b** You will experience it raining sometime in the future.
- **c** You are travelling down an unknown road. The next bend is left.
- **d** A person can walk on water unaided.
- **e** A wine glass will break when you drop it onto the floor.

2 Write down an event that is each of the following.

- **a** Very unlikely
- **b** Certain
- **c** Very likely
- **d** Impossible

3 A child is asked to choose a lucky number from one of the following.

1   2   3   4   5   6   7   8   9

Which of the following is more likely?

- **a** Odd number or number less than 5
- **b** Prime number or even number
- **c** Multiple of 3 or multiple of 4
- **d** Triangle number or cube number

4 Imagine the following quadrilaterals are cut from plastic and placed in a bag.

Rectangle   Parallelogram   Square   Rhombus   Kite   Trapezium

You pick one shape out of the bag at random. Copy and complete the following sentences by filling in the missing probability words.

impossible, very unlikely, unlikely, an even chance, likely, very likely, certain

- **a** Picking a shape with a right angle is ...
- **b** Picking a shape with at least two equal sides is ...
- **c** Picking a shape with four angles is ...

d Picking a shape with two acute angles is …
e Picking a shape with no equal sides is …
f Picking a shape with three sides is …

Practice
3B Probability scales

1 Copy and complete the following table.

Event	Probability of event occurring (p)	Probability of event not occurring ($1 - p$)
A	$\frac{2}{5}$	
B	0.75	
C	$\frac{5}{8}$	
D	0.12	
E	$\frac{19}{20}$	
F	0.875	
G	74%	

2 The probability of an egg having a double yoke is 0.009. What is the probability that an egg does not have a double yoke?

3 100 rings are placed in a box. Twenty are gold, $\frac{1}{4}$ are silver, 16 are plastic and the rest are copper. A ring is chosen at random. What is the probability that it is the following?

a Gold **b** Not silver **c** Copper **d** Not plastic

4 The diagram shows 12 dominoes.

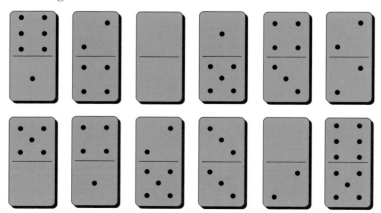

A domino is chosen at random. Calculate the probability that it:

a has a 5 **b** is a double **c** has a total of 7
d is not a double **e** does not have a 5 **f** does not have a blank
g does not have a 3

5 A weather forecaster estimates the probability of rain to be 35%, with a 0.82 probability of black ice, and a 1 in 10 chance of snow. What is the probability of the following?

   a   No snow   b   No black ice   c   No rain

Practice 3C Mutually exclusive events

1 The diagram shows sketches of eight faces.

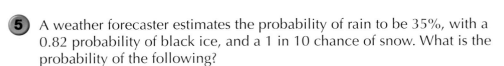

Which of the following pairs of events are mutually exclusive? (Hint: "Left eye" means the eye on the left of the diagram.)

   a   A smiling face. A sad face.
   b   Left eye shut. Both eyes shut.
   c   Wearing a hat. Both eyes open.
   d   Both eyes open. A sad face.
   e   Wearing a hat. Right eye shut.
   f   Smiling with an eye open. Right eye shut.

2 A six-sided die is numbered 1, 2, 3, 4, 5, 6. It is rolled once. Here are some events.

   i   Number 3   ii   Even number   iii   Number greater than 3
   iv   Square number   v   Triangle number   vi   Multiple of 3
   vii Prime number   viii Number 1

   a   Write down three pairs of events that are mutually exclusive.
   b   Write down three pairs of events that are *not* mutually exclusive.

3 Two dice, each numbered 1, 2, 3, 4, 5, 6, are rolled together. Make a list of the possible outcomes, for example, (1, 6).

Practice 3D Calculating probabilities

1 The letters of the word SUCCESSOR are each written on a card and placed in a bag. One of the cards is withdrawn from the bag. What is the probability that the letter is the following?

   a   S   b   A vowel
   c   One of the last 10 letters of the alphabet
   d   C or S   e   A consonant   f   A letter of the word ROSE

2 Two children, Kim and Franz, each write their name on a card and put the card in a bag. The following coins are placed in the same bag: 1p, 2p, 5p, 10p, 20p, 50p, £1. A card and a coin are taken from the bag at random. The coin is given to the named person.

  a Make a table to show the possible outcomes.
  b Calculate the probability of the following.
     i Kim receives 20p
     ii Franz receives less than 10p
     iii One of the children receives 10p
     iv Kim does not receive £1
     v Neither child receives more than 20p

3 A taxi firm owns a red and a green taxi cab. The red taxi can carry up to six passengers. The green taxi can carry up to five passengers. Both taxis are in continuous use, that is, always have at least one passenger.

  a Copy and complete the table showing the total number of passengers being carried at any one time.

		Red taxi			
		1	2	3	
Green taxi	1	2			
	2				
	3			6	

  b What is the probability that, at any one time, the number of passengers being carried is the following?
     i 7
     ii 2
     iii 12
     iv Less than 5
     v An odd number
     vi 2 or 7

3E Experimental probability

1 The numbers of days it rained over different periods are recorded below.

Recording period	Number of days of rain	Experimental probability
30	12	
60	33	
100	42	
200	90	
500	235	

  a Copy and complete the table.
  b What is the best estimate of the probability of it raining? Explain your answer.
  c Estimate the probability of it not raining.
  d Is there a greater chance of it raining or not raining?

2 If you drop a matchbox, it can land in one of three positions.

END

EDGE
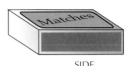
SIDE

a Drop a matchbox 10 times. Record your results in the tally chart below.

How matchbox landed	Tally	Frequency	Experimental probability
End			
Edge			
Side			

b Calculate the experimental probabilities.
c Repeat the experiment. This time drop the matchbox 20 times.
d Repeat the experiment with larger numbers of trials. Stop when you are confident that you have good estimates of the probabilities.

CHAPTER 4 Number 2

Practice

4A Fractions and decimals

1 Write the following decimals as fractions with a denominator of 10, 100 or 1000. Then cancel to their simplest form.

   a 0.6   **b** 0.36   **c** 0.65   **d** 0.255   **e** 0.025

2 Without using a calculator, convert these fractions to decimals.

   a $\frac{13}{50}$ **b** $\frac{17}{25}$ **c** $\frac{4}{5}$ **d** $\frac{3}{20}$

3 Which of the following fractions have recurring decimals? Try to answer the questions without using a calculator. Then check to see if you are correct.

   a $\frac{3}{8}$ **b** $\frac{2}{7}$ **c** $\frac{9}{25}$ **d** $\frac{13}{40}$ **e** $\frac{5}{6}$
   f $\frac{19}{50}$ **g** $\frac{4}{9}$ **h** $\frac{3}{14}$ **i** $\frac{15}{16}$ **j** $\frac{16}{21}$

4 Find the larger of each pair of fractions.

   a $\frac{3}{25}$ and $\frac{1}{7}$ **b** $\frac{11}{18}$ and $\frac{9}{16}$ **c** $\frac{7}{11}$ and $\frac{29}{40}$

5 Write these fractions in order of increasing size.

$\frac{5}{7}, \frac{13}{16}, \frac{33}{50}, \frac{22}{27}$

6 Find fractions that give the calculator displays shown in **a** through **i**. Each denominator is one of the following: 9, 99, 999, 90, 990, 9999, 900, 9900, 99900.

Example: 0.013131313 The recurring digits are 13.

Try $\frac{13}{99}$ 0.131313131 *Wrong*   Try $\frac{13}{990}$ 0.013131313 *Correct*

- **a** 0.888888888
- **b** 0.737373737
- **c** 0.657657657
- **d** 0.123412341
- **e** 0.044444444
- **f** 0.036363636
- **g** 0.020202020
- **h** 0.002828282
- **i** 0.005005005

Practice 4B Adding and subtracting fractions

1 Find the least common multiple of the following pairs of numbers.

- **a** 2, 5
- **b** 6, 8
- **c** 2, 6
- **d** 10, 15
- **e** 9, 12

In questions 2 through 5, convert the fractions to equivalent fractions with a common denominator. Then work out the answer. Cancel your answers and write them as mixed numbers if necessary.

2
- **a** $\frac{2}{5} + \frac{1}{2}$
- **b** $\frac{5}{8} + \frac{1}{12}$
- **c** $\frac{5}{6} + \frac{1}{3}$
- **d** $\frac{4}{7} + \frac{3}{5}$

3
- **a** $\frac{3}{5} - \frac{1}{2}$
- **b** $\frac{5}{9} - \frac{1}{6}$
- **c** $\frac{7}{8} - \frac{2}{3}$
- **d** $\frac{7}{10} - \frac{1}{4}$

4
- **a** $\frac{3}{7} + \frac{2}{3}$
- **b** $\frac{5}{12} + \frac{1}{8}$
- **c** $\frac{7}{9} + \frac{5}{6}$
- **d** $\frac{3}{5} + \frac{1}{2} + \frac{7}{10}$
- **e** $\frac{7}{9} - \frac{1}{2}$
- **f** $\frac{11}{15} - \frac{2}{5}$
- **g** $\frac{5}{6} - \frac{1}{10}$
- **h** $\frac{2}{3} + \frac{5}{6} - \frac{5}{12}$

5
- **a** $2\frac{1}{5} - \frac{7}{10}$
- **b** $1\frac{3}{4} + \frac{5}{6}$
- **c** $3\frac{1}{8} - 2\frac{2}{3}$
- **d** $4\frac{1}{10} - 2\frac{1}{4}$

Practice 4C Multiplying and dividing fractions

1 Use grids to work out the following.

- **a** $\frac{1}{4}$ of 28
- **b** $\frac{3}{5}$ of 30
- **c** $\frac{1}{3}$ of 36
- **d** $\frac{3}{4}$ of 24

2 Calculate the following.

a $\frac{4}{9}$ of 36 kg   b $\frac{5}{12}$ of 600 ml   c $\frac{2}{7}$ of 98 cm   d $\frac{1}{4}$ of 52 km

e $\frac{3}{7}$ of 42 cm   f $\frac{7}{10}$ of 50 grams   g $\frac{3}{5}$ of £80   h $\frac{7}{8}$ of 640 litres

3 Work out the following. Cancel your answers and write them as mixed numbers if necessary.

a $4 \times \frac{3}{7}$ b $9 \times \frac{5}{6}$ c $7 \times \frac{3}{10}$ d $8 \times \frac{3}{4}$

4 Work these out. Cancel your answers and write them as mixed numbers if necessary.

a $\frac{2}{5} \div 3$ b $\frac{8}{9} \div 6$ c $\frac{2}{3} \div 6$ d $\frac{3}{7} \div 4$

Practice 4D Percentages

1 Without using a calculator, express the following.

a 19 as a percentage of 25   b 13 as a percentage of 20
c 96 as a percentage of 400   d 3 as a percentage of 50

2 Use a calculator to find the following, correct to the nearest per cent.

a 14 as a percentage of 23   b 81 as a percentage of 120
c 6 as a percentage of 65   d 3200 as a percentage of 7000

3 Milton drinks 35 cl of an 80 cl bottle of orange.

a What percentage did he drink?
b What percentage remains?

4 A company employs 280 workers. At the last general election, 130 voted Labour, 83 voted Liberal Democrat and the remainder voted Conservative. Calculate the percentage vote for each party.

5 Three people contributed £23, £18.50 and £25.60 toward their total restaurant bill. What percentage of the total bill did each person contribute?

6 a Stuart took a maths test lasting 80 minutes. He spent 23 minutes on the Algebra section, 33 minutes on the Number section and the remaining time on the Geometry section. What percentage of the time did he spend on each section?

b Marcel spent the same amount of time on each section as Stuart. But he also took an Extension section lasting 30 minutes. What percentage of the time did he spend on each section?

Practice 4E Percentage increase and decrease

Do not use a calculator for the first three questions.

1
- **a** Increase $40 by 20%.
- **b** Decrease 200 kg by 5%.
- **c** Decrease 7p by 60%.
- **d** Increase 2000 m by 95%.
- **e** Increase 230 ml by 45%.
- **f** Decrease £7.60 by 85%.

2 Before a typing course, Leon could type 60 words per minute. The typing course increased his speed by 35%. What was his speed after the course?

3 Three people each withdrew a certain percentage of their bank balance from their bank account.

John: 29% of £300   Hans: 75% of £640   Will: 15% of £280

Which person has the largest remaining bank balance?

Use a calculator for the last four questions.

4
- **a** Increase 254 cm^2 by 18%.
- **b** Decrease £15.23 by 84%.
- **c** Increase 7143 g by 47%.
- **d** Increase 0.74 litres by 31%.
- **e** Decrease £17 400 by 97%.
- **f** Decrease 715 kg by 28%.

5 A coat costs £72 before VAT of 17.5% is added. What is the price including VAT?

6 Marvin's average score on the computer game Space Attack was 256.
- **a** After buying a new gamepad, his average score increased by 28%. What was his new average score, to the nearest whole number?
- **b** Marvin went on holiday. When he returned, his average score decreased by 15%. What was his new average score, to the nearest whole number?

7 'Zipping' a computer file reduces its size by a certain percentage. Find the smallest file after zipping.
- **a** 500 kB file reduced by 17%
- **b** 740 kB file reduced by 43%
- **c** 655 kB file reduced by 23%
- **d** 1264 kB file reduced by 92%

Practice 4F Real-life problems

1 Marcia has an annual salary of £24 000. She has a tax allowance of £4800, that is, she pays no tax on that amount. She pays 10% on the first £1880 of her taxable income. She pays 22% tax on the remaining taxable income.
- **a** How much tax does Marcia pay altogether?
- **b** What is her annual salary after paying tax?

c) What is her monthly income after tax?

d) She pays 3% of her salary (after tax) into a pension scheme. How much does she pay each year into her pension, to the nearest penny?

e) Marcia receives a 7% increase in her annual salary. Her tax allowance remains the same. Recalculate the answers to parts **a** to **d**.

2) Mrs Walker left £45 750 in her will. The money was divided between her three children as follows.

    Derek 37%   Maria 29%   Jason 34%

The solicitors deducted 6% for expenses. The remainder was divided amongst the children. How much money did each child receive?

3) Three dealers offer the following repayment options for a car with a marked price of £12 600. Which works out the cheapest overall?

Trustworthy Cars 20% deposit followed by 12 monthly payments of £900
Bargain Autos   114% of the marked price spread over 18 months
Future Car Sales 7% of the marked price for each of the first 6 months followed by 6 monthly payments of £1300

4) Amanda deposits £3000 in her bank. She keeps the money in the bank for five years. The bank pays interest of 4% per year. Calculate the amount of interest Amanda receives after 5 years if she does the following.

a) She spends the interest at the end of each year.

b) She adds the interest to her bank account at the end of each year. (Hint: Work out the amount in her account at the end of Year 1, then Year 2 and so on.)

CHAPTER 5 Algebra **2**

Practice

5A Algebraic shorthand

1) Write each of these expressions using algebraic shorthand.

   a $x \div 5$ **b** $3 \times b$ **c** $4 \times m \times n$
   d $p \div q$ **e** $2 \times (x - 1)$ **f** $5 \times m + 3$
   g $c \div (d + 2)$ **h** $5 \times t \div 4$ **i** $(a + b) \times (m - n)$

2) Simplify the following expressions.

   a $6n \times t$ **b** $5k \times 3m$ **c** $h \times 5g$ **d** $2u \times 3t \times 4v$

3. Which of the following statements are correct and which are incorrect? If you are unsure, check by substituting a number for the letter.

   a $6 \div d$ is the same as $\dfrac{d}{6}$
   b $4(a + 2)$ is the same as $2(a + 4)$
   c $\dfrac{10m}{5}$ is the same as $2m$
   d $\dfrac{12}{3n}$ is the same as $4n$
   e $(a - 2) \div 2$ is the same as $2 \times (a + 2)$
   f $6 \times d \div 2$ is the same as $\dfrac{6d}{2}$

4. Write down the equivalent expressions, for example, $5d = 5 \times d = d5$.

   (Hint: Three of the expressions have no equivalent.)

   $st + 3$ $3 - st$ $3 + st$
   $t \times 3 + s$ $\dfrac{s}{3} + t$
   $3 \div s + t$ $s3t$ $3 + t \times s$
   $3s + t$ $3 \times s \times t$ $t \times s + 3$
   $t + \dfrac{3}{s}$ $t + s3$ $s \times 3 + t$
   $s + 3t$ $ts - 3$

5. Solve each of the following equations. In your solution, write each equation on a separate line.

   a $5x + 7 = 22$
   b $2p - 9 = 7$
   c $10m + 15 = 125$
   d $12c - 13 = 71$
   e $4k + 8 = 19$
   f $8f - 25 = 98$

6. Which of the following statements are correct and which are incorrect?

   a $5 \times p = 12$ is the same as $12 = 5p$
   b $5m + 2 = 17$ is the same as $17 = 5m + 2$
   c $9 = 2j - 4$ is the same as $2j + 4 = 9$
   d $n + 3 = 12$ is the same as $12 = 3 + n$
   e $2t + 3 = 9$ is the same as $3 = 2t + 9$
   f $7 - 3p = 10$ is the same as $10 = 3p - 7$

Practice

5B Like terms

1. Make a list of the terms in each of the following.

   a $y + 2x - 3$
   b $4x = 3 + 2x$
   c $\dfrac{4}{t} - u$
   d $3x^2 = 14x + 2$

   Simplify the following expressions.

2. a $4i + 7i$
   b $7r - 2r$
   c $-u - u$
   d $3h + 2h - h$
   e $6y - 8y$
   f $-3m + 5m$
   g $4t - 3t - 6t$
   h $4n - 3n$
   i $5zt + 4zt$
   j $-ab - ab$
   k $2ad + 7ad - 10ad$
   l $3f^2 - f^2$

3
a $6k + 4k + 3l$
b $7h + 3i - 2i$
c $10y + 2x + 3y$
d $4p + 2p - 1$
e $4d - 7d + 2e$
f $7t - 2u - 4t$
g $9w + 3x - 10w$
h $3fg + 7f + 2fg$
i $s^2 - t^2 + 3s^2$

4
a $4q + 3q + 6i + i$
b $8z - 3z + 4b - 2b$
c $9u + 3v + 2u + 4v$
d $7j + 5k - 3k + 2j$
e $3m - 5m - 4n + 7n$
f $4d + 6e - 8e + 2d$
g $-3c + 4d + 8c - 5d$
h $10r - 4s - 7s - 8r$
i $g - h - 2g + 2h$

Practice
5C Expanding brackets

1 Expand the following brackets.

a $2(f + k)$ b $s(t - 4)$ c $4(2a + 3)$ d $m(n + 3r)$
e $3(3w - 2s)$ f $a(2b - 3)$ g $5(s + t - u)$ h $d(2 - 3r + 4g)$

2 Expand the following brackets.

a $-(p + q)$ b $-(3r - s)$ c $-4(a + b)$ d $-2(m - n)$
e $-5(2d - 3e)$

Expand and simplify the following expressions.

3
a $4(f + g) + 6f$
b $2k + 3(3k - s)$
c $4x + 2(2y - 3x)$

4
a $5(p + q) + 2(2p + 3q)$
b $4(2i + j) + 3(i - j)$
c $2(3b - 2a) + 4(b - 2a)$
d $5(m - 3n) + 2(3m + 4n)$

5
a $4t - (2t + 3u)$
b $7m - (2m - n)$
c $2x - (4y + 3x)$

6
a $4(h + i) - (2h + 3i)$
b $6(2s + t) - (3s - 2t)$
c $5(2w - v) - 3(4w + 2v)$

Practice
5D Using algebra with shapes

1 Write down the perimeter and area of each shape as simply as possible.

a   b 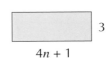   c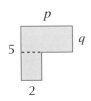

d   **e**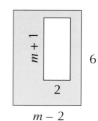

2 a Write down the area of the large parallelogram.
b Write down the areas of the smaller parallelograms, A to D.
c Show that the sum of the smaller areas is the same as your answer to part **a**.

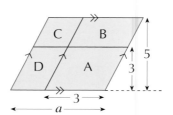

3 For each question, mark any necessary lengths on your diagram.

a Sketch a rectangle whose area is $5a$.
b Sketch a rectangle whose perimeter is $6x + 4$.
c Sketch a triangle whose area is $8x$.
d Sketch a trapezium whose area is $5x$.

Practice
5E Use of index notation with algebra

1 Write the following expressions using index form.

a $g \times g \times g \times g \times g \times g$ **b** $9k \times k$ **c** $5t \times 3t$
d $r \times 4r$ **e** $2j \times 2j \times 2j$

2 Write the following expressions as briefly as possible.

a $m + m + m + m + m + m + m + m$
b $t \times t \times t \times t \times t$ **c** $d + d + d + e \times e \times e$

3 Explain the difference between $6w$ and w^6.

4 Expand the following brackets.

a $v(3 + v)$ **b** $m(3m - 2n)$ **c** $D(3E - 2D)$ **d** $3s(2s + 3t)$

5 Expand and simplify the following expressions.

a $2ab + a(3b + 2)$ **b** $v(2v + 4t) - 2vt$ **c** $9qz - q(3z - 2q)$

6 Expand and simplify the following expressions.

a $r(2r + s) + s(3r + s)$ **b** $p(3p - 2q) + q(4p - 5q)$
c $m(3m + 2n) - n(2m + 4n)$ **d** $c(3c - 2d) - d(5c - d)$

7 Expand and simplify the following expressions.

a $2g^2 + g(3g + 4)$ **b** $d(2d + 4e) + d(3d - e)$
c $w(4 - 2w) - w(2w + 3)$

CHAPTER 6
Geometry and Measures 2

Practice
6A Area of a triangle

1 Calculate the areas of the following triangles.

a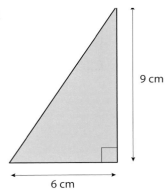

b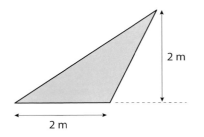

c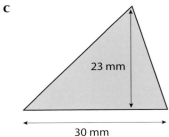

d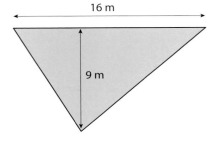

e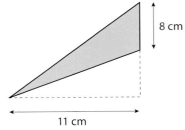

f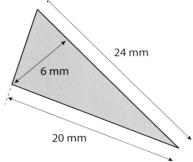

2 Copy and complete the table below which gives the measurements of five triangles.

Base	Height	Area
12 cm	9 cm	
8 cm	14 cm	
6 mm	7 mm	
16 cm		64 cm^2
	20 m	100 m^2

3. Use squared paper to draw four different triangles with area 24 cm².

   (Hint: Draw the base first, using a whole number of centimetres that is a factor of 48, for example, 8 cm. Then calculate the height of the triangle.)

4. Find the area of each puzzle piece. Each piece is a compound shape. Each small square represents a square centimetre. Show your calculations.

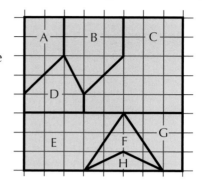

Practice
6B Area of a parallelogram

1. Calculate the areas of the following parallelograms.

   a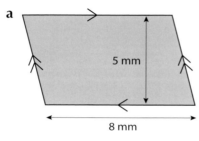

   b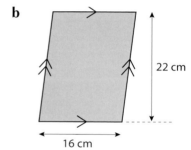

   c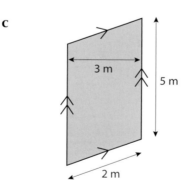

   d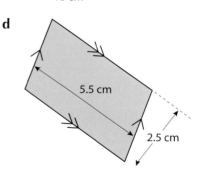

2. Copy and complete the table below which gives the measurements of five parallelograms.

Base	Height	Area
7 cm	13 cm	
9 m	19 m	
250 mm	70 mm	
	15 m	120 m²
12 cm		30 cm²

(3) Use squared paper to draw four different parallelograms with area 48 cm².

(Hint: Draw the base first, using a whole number of centimetres that is a factor of 48, for example, 8 cm. Then calculate the height of the parallelogram.)

(4) Find the area of each puzzle piece. Each piece is a compound shape or a parallelogram. Each small square represents a square centimetre. Show your calculations.

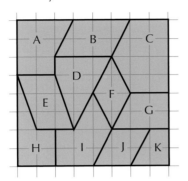

Practice
6C Area of a trapezium

(1) Calculate the areas of the trapezia below.

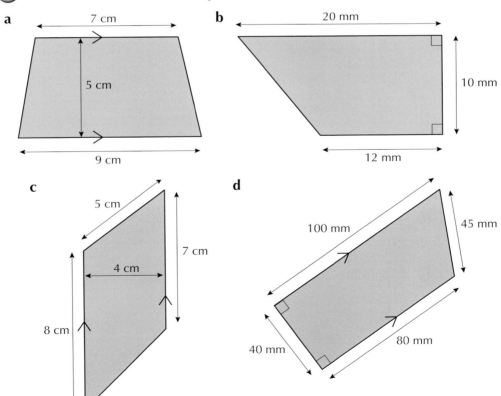

2) Copy and complete the table below for trapezia **a** through **e**.

Trapezium	Parallel side a	Parallel side b	Height h	Area
a	7 cm	9 cm	3 cm	
b	13 m	8 m	5 m	
c	2 mm	6 mm		32 mm²
d		4 m	6 m	60 m²
e	12 cm		10 cm	250 cm²

3) Use squared paper to draw four different trapezia with area 24 cm².

(Hint: Decide the lengths of the parallel lines first and make sure that their total length is a factor of 48. For example, parallel sides 7 cm and 5 cm have a total length of 12 cm, which is a factor of 48. Then calculate the height of the trapezium.)

4) Find the area of each puzzle piece. Each piece is a compound shape or a trapezium. Each small square represents a square centimetre. Show your calculations.

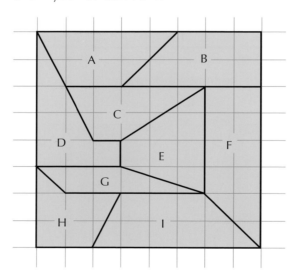

Practice

6D Volume of a cuboid

1) **a** Find the surface area of each of the following cuboids.

i

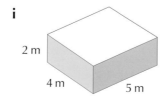

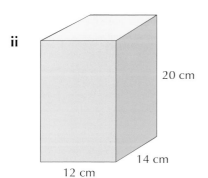

ii

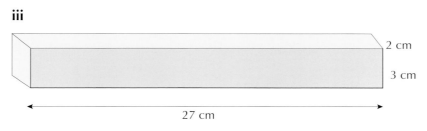

iii

b Find the volume of each of the above cuboids.
c Find the capacity of each of the above cuboids. Use suitable units of capacity.

2 Calculate the volume, surface area and capacity of a cube of side 6 cm.

3 Krispies are sold in three sizes: mini, medium and giant. The boxes are filled to the top.

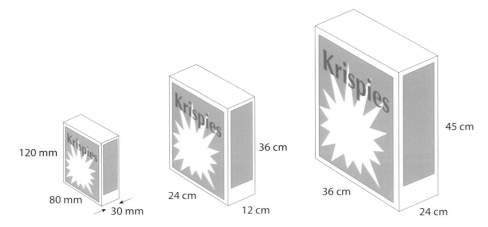

a Calculate the volume of each box.
(Hint: Convert mm to cm.)
b How many times bigger is the giant box compared to the medium box?
c How many mini boxes could be fitted into the following?
   i A medium box   **ii** A giant box
d If 20 cm^3 of Krispies weigh 1 gram, calculate the weight of Krispies contained in each box, to the nearest gram.

4 Calculate the volume of each of the following 3-D shapes.

a

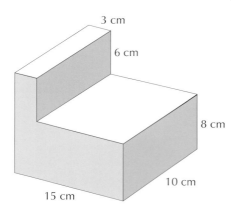

b

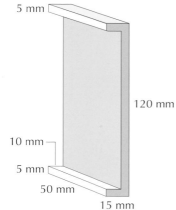

Practice
6E Imperial units

1 Convert each of the following quantities to the units shown in brackets.
- **a** 25 st (lb)
- **b** 7 miles (yd)
- **c** 9 ft 7 in (in)
- **d** $3\frac{1}{2}$ ton (lb)

2 Convert each of the following quantities to the units shown in brackets.
- **a** 83 oz (lb and oz)
- **b** 532 ft (yd and ft)
- **c** 75 pt (gall and pt)
- **d** 125 in (ft and in)

3 a How many stones are in a ton? **b** How many feet are in a mile?

4 Convert each Imperial quantity to the approximate metric quantity shown in brackets.
- **a** 11 oz (g)
- **b** 270 miles (km)
- **c** 100 in (m)
- **d** $7\frac{1}{4}$ lb (kg)

5 Convert each metric quantity to the approximate Imperial quantity shown in brackets.
- **a** 72 km (miles)
- **b** 420 g (oz)
- **c** 81 litres (gall and pt)
- **d** 720 g (lb and oz)

6 Calculate the approximate length of this tape measure in yards, feet and inches.

Use the approximation 1 in ≈ 2.5 cm.

7 A small Dutch beer glass holds 0.25 litres of beer. Approximately how many pints does it hold?

 8 Convert this recipe to Imperial units, using appropriate approximations.

Biscuits

750 g plain flour
360 g butter
315 g sugar
75 g rice flour
15 g salt

 9 You know that 1 in ≈ 2.5 cm and 1 oz ≈ 30 g.
Find a close metric approximation (shown in brackets) for each of the following measurements.

    a 1 yd (cm)   **b** 1 lb (g)   **c** 1 mile (m)

CHAPTER 7 Algebra 3

Practice

7A Linear functions

1 a Copy the mapping diagram for the function $x \to x - 3$.

  b Map the integer values from $x = -1$ to $x = 4$.
  c Map the following values.
    i $x = 3.5$ **ii** $x = 0.5$ **iii** $x = 1.5$ **iv** $x = -0.5$

2 a For each of the following functions, draw a mapping diagram using two number lines from $x = -5$ to $x = 10$. Label your diagrams. Map all the integer values that will fit onto the diagram.
    i $x \to x + 3$ **ii** $x \to x - 1$ **iii** $x \to 2x$
    iv $x \to 2x + 4$ **v** $x \to 2x - 2$ **vi** $x \to 3x + 2$

  b Map the following values.
    i $x = 2.5$ **ii** $x = 0.5$ **iii** $x = -1.5$ **iv** $x = -0.5$

3 Draw a mapping diagram for the function $x \to -2x$ using two number lines from $x = -6$ to $x = 6$. Map all the integer values that will fit onto the diagram.

Practice 7B Finding a function from its inputs and outputs

Find the function that maps the following inputs to the outputs.

1
a $\{1, 2, 3, 4, 5\} \rightarrow \{7, 8, 9, 10, 11\}$
b $\{3, 4, 5, 6, 7\} \rightarrow \{6, 8, 10, 12, 14\}$
c $\{0, 1, 2, 3, 4\} \rightarrow \{-2, -1, 0, 1, 2\}$
d $\{1, 2, 3, 4, 5\} \rightarrow \{3, 5, 7, 9, 11\}$
e $\{2, 3, 4, 5, 6\} \rightarrow \{5, 8, 11, 14, 17\}$
f $\{-2, -1, 0, 1, 2\} \rightarrow \{1, 3, 5, 7, 9\}$
g $\{-1, 0, 1, 2, 3\} \rightarrow \{-6, -2, 2, 6, 10\}$

2
a $\{3, 4, 5, 8, 9\} \rightarrow \{8, 11, 14, 23, 26\}$
b $\{2, 4, 5, 6, 8\} \rightarrow \{8, 14, 17, 20, 26\}$
c $\{0, 4, 6, 7, 8\} \rightarrow \{-3, 17, 27, 32, 37\}$
d $\{1, 3, 5, 7, 9\} \rightarrow \{4, 12, 20, 28, 36\}$
e $\{3, 6, 9, 12, 15\} \rightarrow \{3, 9, 15, 21, 27\}$

3
a $\{1, 2\} \rightarrow \{5, 7\}$ b $\{2, 4\} \rightarrow \{4, 10\}$ c $\{0, 3\} \rightarrow \{1, 16\}$

Practice 7C Graphs from functions

1 a Copy and complete the table below for the function $y = 3x$.

x	−3	−2	−1	0	1	2	3
y = 3x							

b Draw a grid with its x-axis from −3 to 3 and y-axis from −10 to 10.
c Draw the graph of the function $y = 3x$.

2 a Copy and complete the table below for the function $y = 4x - 3$.

x	0	1	2	3	4	5
y = 4x − 3						

b Draw a grid with its x-axis from 0 to 5 and y-axis from −5 to 20.
c Draw the graph of the function $y = 4x - 3$.

3 a Copy and complete the table for each of the functions.

x	−2	−1	0	1	2	3
y = x + 2						
y = 2x + 2						
y = 3x + 2						
y = 4x + 2						

b Draw a grid with its x-axis from −2 to 3 and y-axis from −10 to 15.

c Draw the graph of each function in the table.
d What is the same about the lines?
e What is different about the lines?
f Use a dotted line to sketch the graph of $y = 2.5x + 2$.

Practice
7D Gradient of a straight line (steepness)

1 a Write down the gradient of each of the following lines.
b Write down where each line crosses the y-axis.

i 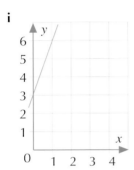 ii 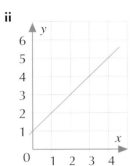 iii 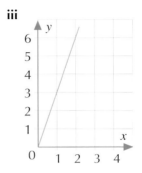 iv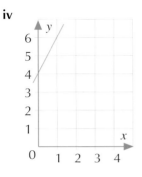

2 Write down the equation of each of the lines in question 1.

3 a Plot the points $A(1, 3)$ and $B(3, 5)$.
b Work out the gradient of AB.
c Extend the line to cross the y-axis.
d Write down the equation of the line that passes through AB.

Practice
7E Real-life graphs

1 a Copy the grid below.

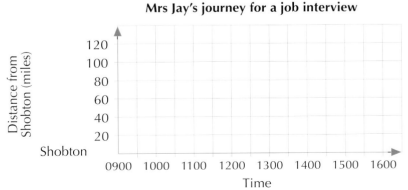

b Draw on the grid a distance–time graph for the following train journey. Mark the places Mrs Jay visited on the vertical axis.

Mrs Jay had to travel to a job interview at Penford. She caught the 0900 train from Shobton and travelled to Deely, 30 miles away, in an hour. There she had to wait 30 minutes for a connecting train to Penford. This part of the journey took $1\frac{1}{2}$ hours at a speed of 60 mph. Her job interview at Penford lasted 2 hours. Her return journey lasted $1\frac{1}{2}$ hours.

c Calculate the average speed of the train on her return journey.

2 a Draw a grid with the following scales.

- Horizontal axis represents time, from 0 to 6 hours, 1 cm to 30 minutes.
- Vertical axis represents distance from factory, from 0 to 50 miles, 1 cm to 5 miles.

b Draw on the grid a distance–time graph for the following journey. Mark the places of delivery on the vertical axis.

A car transporter left the factory and took 30 minutes to travel 15 miles to a dealer in Harton. It took half an hour to unload three of the cars. The transporter travelled at 25 mph for another hour and made a delivery at Glimp. This delivery and lunch took an hour. A final delivery was made 30 minutes later at Unwich after a 10-mile drive. This delivery took 30 minutes. The transporter returned to the factory, taking two hours to get back.

c Calculate the average speed of the transporter:
   i between Glimp and Unwich   **ii** on the return journey.

3 a Draw a grid with the following scales.

- Horizontal axis represents time, from 0 to 30 minutes, 1 cm to 2 minutes.
- Vertical axis represents temperature, from 20 °C to 100 °C, 1 cm to 10 °C.

b Draw a graph on the grid to show the temperature of Stan's CupSoup.

Stan had a mug of CupSoup at 20 °C. His microwave took 4 minutes to boil the CupSoup at full power. He forgot about the soup for 9 minutes, by which time it had cooled to 80 °C. So he heated it to boiling point on quarter power. It took him 9 minutes to eat the soup. The temperature of the last spoonful was 50 °C.

CHAPTER 8 Number 3

Practice

8A Powers of 10

Do not use your calculator.

1 Multiply each of the following numbers by 10^4.
   a 2.7   b 0.05   c 38   d 0.008

2 Divide each of the following numbers by 10^3.
   a 730   b 4   c 2.8   d 35 842

3 Calculate the following.
   a 7.4×10^3 b 13×10^5 c $0.87 \div 10^3$
   d $17.4 \div 10^2$ e 0.0065×10^3 f $19.4 \div 10^4$

4 Multiply the following numbers by:
   i 0.01   ii 0.001
   a 280   b 0.6   c 1.05   d 9851

5 Divide the following numbers by:
   i 0.01   ii 0.001
   a 4   b 800   c 0.9   d 67.2

6 Work your way along this chain of calculations for each of the following starting numbers.
   a 2000   b 7   c 0.6

7 Round the following numbers to:
   i one decimal place
   ii two decimal places
   a 8.265   b 3.965   c 0.047   d 4.994   e 0.095

8 Work out the following.
   a 14.281×10 b 6.3942×10^2 c $225.61 \div 10$

Round your answers to one decimal place.

Practice
8B Large numbers

1 Write the following numbers in words.
  a 956 348   b 15 230 421   c 8 002 040   d 604 500 002

2 Write the following numbers using figures.
  a Two hundred and six thousand, one hundred and seven
  b Fifty million, thirty-two thousand and eight

3 The graph shows the numbers of shares sold every hour during a trading day. Estimate the number sold each hour. Make a table for your answers.

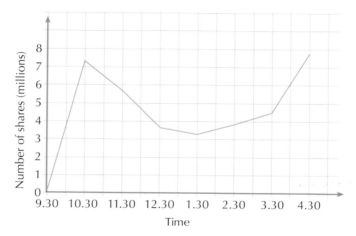

4 Round the following numbers to the nearest:
  i ten thousand   ii hundred thousand   iii million
  a 7 247 964   b 1 952 599   c 645 491   d 9 595 902

5 The top six UK airports in 2006, measured by number of passengers, are shown in the following table.

1	Heathrow	67 339 000
2	Gatwick	34 080 000
3	Stansted	23 680 000
4	Manchester	22 124 000
5	Luton	9 415 000
6	Birmingham	9 056 000

  a Which two of these airports had the same number of passengers, to the nearest million?
  b How many passengers used Heathrow, Gatwick or Stansted in 2006? Give your answer to the nearest ten million passengers.

Practice 8C Estimations

1 Explain why these calculations must be wrong.
   a 53 × 21 = 1111   b 58 × 34 = 2972   c 904 ÷ 14 = 36

2 Estimate answers to each of these problems.
   a 6832 − 496   b 28 × 123   c 521 ÷ 18
   d 770 × 770   e $\dfrac{58.9 + 36.4}{22.5}$

3 Which is the best estimate for 15.4 × 21.6?
   a 16 × 22   b 15 × 21   c 15 × 22   d 16 × 21

4 a Football socks cost £3.71 a pair. Without working out the correct answer, could Ian buy 5 pairs using a £20 note? Explain your answer.
   b Shoelaces cost 68p a pair. The shopkeeper charged Ian £3.25 for 5 pairs. Without working out the correct answer, explain why is this incorrect.

5 Estimate the number to which each arrow is pointing.

   a

   b

   c

Practice 8D Adding and subtracting decimals

Do not use a calculator. Show your working.

1 Calculate the following.
   a 2.06 + 9.77 + 12.3   b 0.87 + 1.79 − 0.94
   c 78.008 − 23.7 − 19.08   d 9.231 − 2.076 − 1.8
   e 13 + 91.03 − 2.378 − 18.26 + 33.333

2 Calculate the following. Work in metres.
   a 5 m − 2.56 m + 108 cm   b 0.95 m + 239 cm − 1.86 m
   c 6 cm + 0.67 m − 0.085 m   d 23.6 cm + 0.082 m − 7.41 cm

3 **a** Calculate the total volume of juice in these full bottles. Work in litres.
(Hint: 1 litre = 100 cl = 1000 ml.)

b All the juice is made into a fruit cocktail. Three cups are drunk. If a cup holds 15.3 cl, how much fruit cocktail is left?

4 The skin is the largest organ in the body and weighs 10.886 kg, on average. The four other largest organs are the liver (1.56 kg), the brain (1.408 kg), the lungs (together weighing 1.09 kg) and the heart (0.315 kg).

a Calculate the total weight of these organs.
b How much more does the skin weigh than the other organs combined?

5 A candle is 32.7 cm tall. It burns down 18.4 mm each day. What is its height at the end of 4 days? Work in centimetres.

Practice
8E Efficient calculations

1 Use the bracket and/or memory keys on your calculator to calculate the following.

a $19.3 - (32.5 - 24.8)$ **b** $(0.24 + 1.73)^2$ **c** $\dfrac{1}{2.09 - 1.93}$

2 Use the fraction key on your calculator to calculate the following.

a $\frac{7}{9} - \frac{1}{6}$ **b** $2\frac{3}{5} + 7\frac{1}{4}$ **c** $9\frac{3}{10} \times 2\frac{1}{2}$

d $1\frac{3}{8} \div (\frac{5}{6} - \frac{2}{9})$ **e** $(5\frac{1}{4} - 2\frac{1}{7})^2$ **f** $\dfrac{3\frac{5}{8} - 2\frac{2}{3}}{4\frac{1}{6} + 3\frac{1}{2}}$

3 Use the power, cube and cube root keys on your calculator to calculate the following. Round your answers to 1 decimal place if necessary.

a 3^7 **b** $\sqrt[3]{29}$ **c** $9.8^2 \times 2.5$

d $14^2 \div 15^2$ **e** $\sqrt{1 + 1.8^3}$

4 Solve each of these problems using the fraction key of your calculator.

   a A wheel turns $17\frac{1}{4}$ turns clockwise, then $3\frac{2}{5}$ turns anticlockwise, then $7\frac{2}{3}$ turns clockwise. How far has it turned altogether?

   b A cat eats $\frac{2}{5}$ of a tin of cat food every day. How long would $5\frac{1}{2}$ tins last?

   c Salami costs £8.58 per kilogram. How much does $3\frac{3}{8}$ kg cost?

   d A giant company Christmas cake is cut into six equal parts. Two fifths of one part is divided between eight office staff.

      i What fraction of the cake does each person receive?

   The remaining cake is divided between 25 factory workers.

      ii What fraction of the cake does each factory worker receive?

      iii What is the difference between the amount received by an office staff member and a factory worker?

Practice
8F Multiplying and dividing decimals

Do not use a calculator.

1 Calculate the following.

   a 5.2×9.3 **b** 3.8×6.6 **c** 4.31×2.7 **d** 0.49×0.78

2 Calculate the following.

   a $2.16 \div 0.4$ **b** $0.24 \div 1.6$ **c** $29.9 \div 23$ **d** $1.12 \div 0.35$

3 Bricks are 23.7 cm long. How far would 17 bricks reach laid end to end?

4 The table shows the postage needed for letters.

Format	Weight	First class Price	Second class Price
Letter	0–100 g	£0.34	£0.24
Large letter	0–100 g	£0.48	£0.40
	101–250 g	£0.70	£0.60

   a What is the cost of sending 14 letters each weighing 90 g by first class mail?

   b How many second class 45 g letters could be posted for £4.80?

   c Which is cheaper to post: 18 first class letters each weighing 130 g, or 26 second class letters each weighing 85 g?

Chapter 9 Geometry and Measures 3

Practice

9A Congruent shapes

1 Use your ruler to check which of these triangles are congruent. Write your answer like this, for example, P = Q.

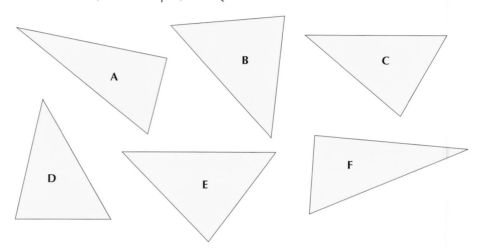

2 Which of these shapes are congruent? Write your answer like this, for example, P = T = W.

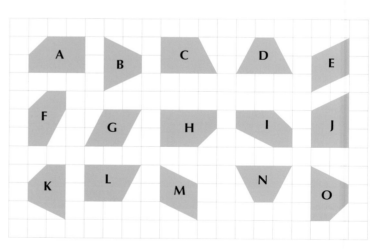

3 a Using triangular dotted paper, draw 15 different shapes by joining some of the dots. Two examples are shown in the diagram below. Label your shapes from A to N.

b Write down the shapes that are congruent.

Practice 9B Combinations of transformations

1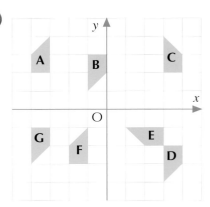

a Describe the single transformation that maps the following.
  i A onto F   ii D onto A   iii C onto A
  iv E onto F   v D onto B
b Describe the combination of two transformations that maps the following.
  i A onto D   ii B onto C   iii E onto G

2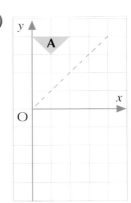

a Copy the diagram onto square paper.
b Reflect triangle A in the dotted mirror line and label the image B.
c Reflect triangle B in the x-axis and label the image C.
d What single transformation maps triangle A onto triangle C?

3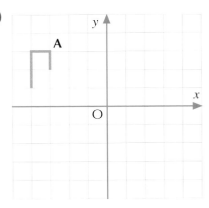

a Reflect shape A in the y-axis and label the image B.
b What combination of two transformations maps shape A onto shape B?

4 a Which single transformation maps a shape onto itself?
b Describe a combination of two transformations that map a shape onto itself.

Practice 9C Enlargements

1. Trace each shape with its centre of enlargement O. Enlarge the shape by the given scale factor.

O ×   Scale factor 3   Scale factor 2

2.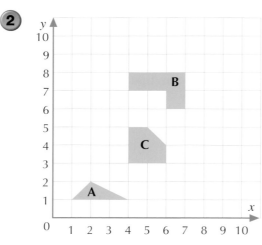

   a Copy the grid and shape A only. Enlarge shape A by a scale factor of 2, using the origin as centre of enlargement. Label the image A'.
   b Copy the grid and shape B only. Enlarge shape B by a scale factor of 2, using the point (7, 10) as centre of enlargement. Label the image B'.
   c Copy the grid and shape C only. Enlarge the shape by a scale factor of 3, using the point (5, 4) as centre of enlargement. Label the image C'.

3.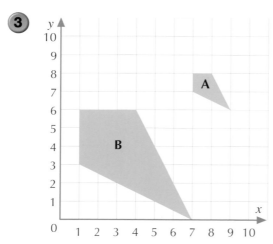

   a Copy the diagram.
   b Shape B is an enlargement of shape A. What is the scale factor?
   c Draw ray lines to find the centre of enlargement. Write down the coordinates of this point.

Practice: 9D Shape and ratio

1 Express each of these ratios in its simplest form.

- **a** 90 cm : 20 cm
- **b** 32 mm : 72 mm
- **c** 150 cm : 2 m
- **d** 1.8 cm : 40 mm
- **e** 0.65 km : 800 m

2

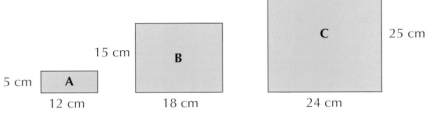

- **a** Find the ratio of the base of rectangle A to the base of rectangle B.
- **b** Find the ratio of the area of rectangle A to the area of rectangle B.
- **c** What fraction is area A of area B?
- **d** Find the ratio of the area of rectangle B to the area of rectangle C.
- **e** Which is greater: area A as a fraction of area B, or area B as a fraction of area C?

3 The diagram shows the design of a new flag.

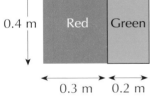

- **a** Calculate the ratio of the red area to the green area.

Use ratios to answer the following questions.

- **b** 60 m² of red cloth is used to make some flags. How much green cloth was needed?
- **c** The total area of some flags is 200 m².
  - **i** How much red cloth do they contain?
  - **ii** How much green cloth do they contain?

4

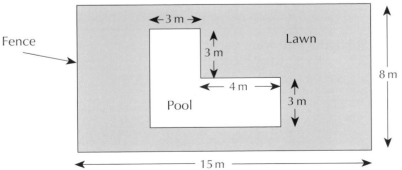

The diagram shows the plan of a garden.

- **a** Find the ratio of the perimeter of the fence to the perimeter of the pool.
- **b** Calculate the area of the pool.
- **c** Calculate the area of the lawn.
- **d** Find the ratio of the lawn area to the pool area.

CHAPTER 10 Algebra 4

Practice 10A Solving equations

1 Solve the following equations.

- **a** $3a + 7 = 13$
- **b** $2n - 3 = 11$
- **c** $5v - 20 = 30$
- **d** $6x + 5 = 29$
- **e** $10c - 17 = 73$
- **f** $2y + 13 = 19$

2 Solve the following equations.

- **a** $8 + 3t = 23$
- **b** $6 + 5m = 16$
- **c** $9 + 4d = 29$
- **d** $13 + 8s = 101$
- **e** $43 + 2z = 51$
- **f** $7 + 7i = 63$

3 Solve the following equations. The answers are decimals or fractions.

- **a** $4q + 1 = 15$
- **b** $5h - 3 = 3$
- **c** $10y + 21 = 69$
- **d** $3f - 2 = 5$
- **e** $8e + 5 = 19$
- **f** $5w - 9 = 5$

4 Solve the following equations. Expand the brackets first.

- **a** $2(n + 1) = 12$
- **b** $5(p - 2) = 25$
- **c** $4(g + 3) = 44$
- **d** $3(2r + 1) = 15$
- **e** $2(3b - 7) = 4$
- **f** $5(4i + 2) = 70$

5 Solve the following equations.

- **a** $4r - 20 = 8$
- **b** $5x + 2 = 9$
- **c** $4(y + 4) = 24$
- **d** $3 + 5a = 48$
- **e** $3(5h - 8) = 36$
- **f** $7 + 4u = 25$

Practice 10B Equations involving negative numbers

Solve the following equations.

1
- **a** $3x + 11 = 2$
- **b** $2y + 15 = 3$
- **c** $4a + 6 = 2$
- **d** $5C + 40 = 15$

2
- **a** $12 + 3x = 6$
- **b** $6 + 5b = 31$
- **c** $22 + 6m = 4$
- **d** $100 + 16q = 36$

3
- **a** $4d - 19 = -3$
- **b** $2x + 7 = -3$
- **c** $3z - 14 = -32$
- **d** $5t - 12 = -22$

4
- **a** $-4i = 28$
- **b** $-3H = 6$
- **c** $-8k = -16$
- **d** $-6x = 30$

5
a $6 - 4x = 18$
b $1 - 7u = 15$
c $3 - 3d = -12$
d $8 - 2m = 16$

6 Expand the brackets first.
a $4(s + 2) = 4$
b $3(m - 5) = -6$
c $2(3n + 9) = 6$
d $5(2y - 12) = -10$

7
a $-5x = 10$
b $8 - 5x = -12$
c $12 + 5f = -3$
d $2(x + 3) = -10$
e $3w - 7 = -1$
f $4(2d - 1) = -36$

Practice
10C Equations with unknowns on both sides

Solve the following equations.

1
a $5y = 9 + 2y$
b $10x = 20 + 6x$
c $8u = 3u + 25$
d $7p = 3p + 32$

2
a $20 - 3x = x$
b $12 - 2c = 2c$
c $4d = 30 - d$
d $5p = 14 - 2p$

3
a $6g + 5 = 2g + 13$
b $9i + 7 = 5i + 19$
c $10h - 3 = 3h + 18$
d $8t - 15 = 6t + 3$

4
a $35 + 2k = 9k$
b $3x + 8 = 5x$
c $2x + 23 = 7x + 3$
d $r + 16 = 5r - 10$

5 These equations involve negative numbers.
a $8y = 6y - 10$
b $6k - 6 = 9k$
c $40 + 7j = 2j$
d $5d + 9 = 3d + 3$
e $-4r = 3r + 21$
f $7n - 3 = 3n - 15$

6 Expand the brackets first.
a $3(x + 2) = 2x + 8$
b $3(2g + 3) = 4g + 17$
c $5(s - 2) = 3(s + 4)$
d $8w - 12 = 3(3w - 1)$

Practice
10D Substituting into expressions

1 Write down the value of each expression.
a $3 + 4x$ when   i $x = 2$ ii $x = 7$ iii $x = -3$
b $9p - 2$ when   i $p = -4$ ii $p = 0$ iii $p = 8$
c d^2 when   i $d = 7$ ii $d = -6$ iii $d = -1$

d $2s^2$ when i $s = 3$ ii $s = 10$ iii $s = -2$
   e $m^2 - 5$ when i $m = 4$ ii $m = 1$ iii $m = -4$
   f $5(3n - 2)$ when i $n = 2$ ii $n = 0$ iii $n = -5$

2 If $p = 3$ and $q = 5$, find the value of each of the following.
   a $2q - p$ b $q + 3p$ c $p^2 + q^2$ d $4(3p - q)$

3 If $r = -2$ and $s = 3$, find the value of each of the following.
   a $2r + s$ b $r - 2s$ c $3(2r + 3s)$

4 If $m = 2$ and $n = -4$, find the value of each of the following.
   a $n^2 - m^2$ b $5m^2$ c mn^2 d $5mn - m^2$

5 If $x = 3$, $y = 5$ and $z = -2$, find the value of each of the following.
   a xyz b $xz + y$ c $2x - 3y - 4z$ d $(x - z)(x + y)$

Practice

10E Substituting into formulae

1 The ideal weight, W kg, of a man of height h cm is given by the formula:
   $W = 0.75h - 67.5$
   Calculate the ideal weight of a man of the following height.
   a 160 cm   b 186 cm   c 1.7 m

2 Given that $C = F + 4d$, find the value of C when:
   a $F = 12, d = 6$ b $F = 200, d = 65$ c $F = 18, d = -3$

3 The cost of a sheet of glass is given by the formula:
   $C = 3wh$
   where C is the cost (£), w the width and h the height in metres. Calculate the cost of the following sheets of glass.
   a Width 2 m, height 4 m   b Width 3 m, height 1.5 m
   c Width 0.3 m, height 0.8 m

4 Given that $W = \dfrac{2m + 2n + p}{5}$, calculate the value of W when:
   a $m = 4, n = 2, p = 6$ b $m = -3, n = 2, p = 5$
   c $m = 6, n = -1, p = -4$

5) The approximate area of a circle of radius r is given by the formula:

$A = 3r^2$

where A is the area and r is the radius.
Find the area of a circle of the following radius.

  a 5 cm   **b** 2.5 cm   **c** 0.9 cm

Practice

10F Creating your own expressions and formulae

1) Write an expression for each of the following using the letters suggested.
   a The sum of the numbers d and 3.
   b The number u reduced by 5.
   c The product of the numbers w, 7 and s.
   d One third of the number m.
   e The number of toes on f feet.

2) **a** How many months are there in Y years?
     b How many metres are there in x centimetres?

3) **a** Michael is H cm tall now. He was 5 cm shorter a year ago. How tall was he then?
     b In three years' time, he will be x cm taller. How tall will he be then?
     c Michael's taller sister Briony has a height of S cm. How much taller than Michael is Briony?
     d What is the average height of Michael and Briony?

4) **a** A Chocolate Wheel costs c pence and a Frother costs f pence. What is the total cost of the following?
       i A Chocolate Wheel and a Frother
      ii Five Frothers
     iii Three Chocolate Wheels and two Frothers
     b How much change from a £2 coin would you receive for each of the purchases in part **a**? Work in pence.

CHAPTER 11 Statistics 2

Practice

11A Stem-and-leaf diagrams

1 The speeds of 35 cars are shown in the stem-and-leaf diagram below.

```
1 | 0 9 9 9
2 | 1 2 2 3 4 4 5 6 6 6 8 9
3 | 0 4 4 5 5 6 6 7 8 8 8 9
4 | 0 3 3 6 8 9
```

Key: 4 | 3 means 43 mph.

a What is the mode? **b** Find the median.
c Calculate the range.
d The speed limit is 30mph. How many cars broke the speed limit?

2 The ages of 23 children are shown in the stem-and-leaf diagram below.

```
11 | 9 9 9 10 11 11
12 | 0 0 1 2 2 3 4 5 7 10 10 10
13 | 0 3 3 4 5
```

Key: 12 | 4 means 12 years and 4 months.

Ages are rounded down to the nearest month.

a How many children are 12 years old?
b How many children are younger than $12\frac{1}{2}$?
c Calculate the mode, median and range.

3 The weights in grams of 27 letters are shown below.

```
48   29   62   48   96   45   30   59   16
88   63   10   74   64   38   90   25    8
44   92   67   78   33   50   23   60   18
```

a Make a stem-and-leaf diagram for the data. Remember to show the key.
b State the mode.
c Calculate the range.
d Find the median.
e Letters weighing 60 g or more are charged 41p for first-class delivery. How many letters need this postage?

Practice 11B Pie charts

1 600 pupils in a school were asked to vote for their favourite soap opera.

The pie chart illustrates their responses.

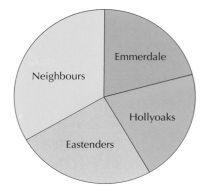

How many voted for each of the following?

a   EastEnders   b   Emmerdale
c   Hollyoaks   d   Neighbours

2 One week on a train, the staff sold 1200 drinks.

The pie chart illustrates the different drinks that were sold.

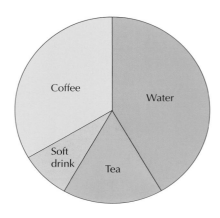

How many of the following drinks were sold that weekend?

a   Water   b   Tea
c   Soft drinks d   Coffee

3 The pie chart illustrates how the 24 absences a class had in one week were distributed on the different days of the week.

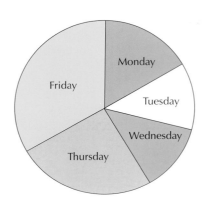

How many pupils were absent on each of the following days?

a   Monday   b   Tuesday
c   Wednesday   d   Thursday   e   Friday

4 Mona carried out a survey on how often they ate chocolate. She asked 72 people.

The pie chart illustrates her results.

How many of these people did the following?

a Never ate chocolate
b Ate chocolate once a week
c Ate chocolate occasionally
d Ate chocolate every day

Practice 11C More about pie charts

1 Draw pie charts to represent the following data.

a The numbers of birds spotted on a field trip.

Bird	Crow	Thrush	Starling	Magpie	Other
Frequency	19	12	8	2	19

b The sizes of dresses sold in a shop during one week.

Size	8	10	12	14	16	18
Frequency	3	7	10	12	6	2

2 150 children went on one of four school summer holidays. How many children chose the following holidays?

a Camping
b France
c Pony trek
d Disney World

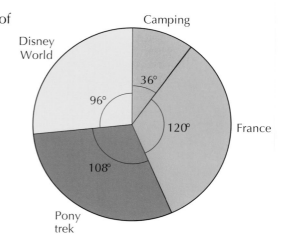

Practice — 11D Scatter graphs

1) After a study into the number of goals scored in their first full season by various prices of footballers and their ages, the following scatter diagrams were created.

Describe the type of correlation and what each graph tells you.

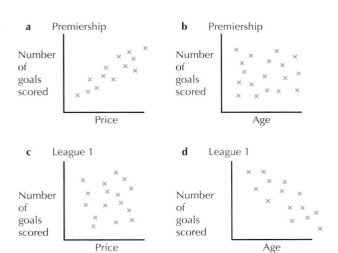

2) After a study into various books and their number of pages, number of chapters as well as the price, the following scatter diagrams were created. Describe the type of correlation and what each graph tells you.

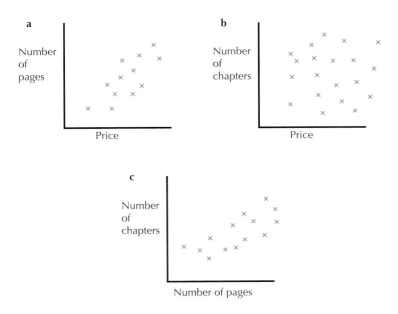

3) After a study into people's ages, weight, height and the average hours of sleep they get, the scatter diagrams on page 50 were created.
Describe the type of correlation and what each graph tells you.

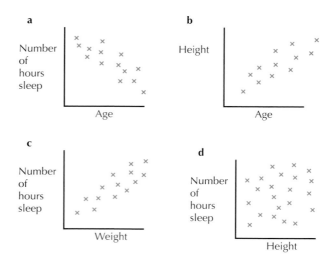

Practice 11E More about scatter graphs

1 The table shows the positions pupils have in their Science and Mathematics classes.

Pupil	Jo	Ken	Lim	Tony	Dee	Sam	Pat	Les	Kay	Val	Rod
Science	24	3	13	21	5	15	22	7	12	27	14
Maths	20	5	10	14	9	15	26	1	16	24	11

a Draw a scatter graph for the data. Use the x-axis for Science position, from 0 to 30. Use the y-axis for Mathematics position, from 0 to 30.
b Describe what the graph tells you.
c Which of the following best describes the degree of correlation: positive correlation, negative correlation, no correlation?

2 The table shows the heights of 20 children and their scores on a History test.

Height (cm)	130	110	160	120	120	140	160	150	110	130	140	120	170	150	140	160	110	150	140	140
Score	13	19	12	14	7	20	14	10	9	3	17	20	20	15	8	6	13	20	6	12

a Draw a scatter graph for the data. Use the x-axis for height, from 100 cm to 180 cm. Use the y-axis for History test score, from 0 to 20.
b Describe what the graph tells you.
c Describe the degree of correlation.

CHAPTER 12 Number 4

Practice

12A Fractions

1 Copy and complete the following.

a $\dfrac{9}{4} = \dfrac{\square}{12}$ b $\dfrac{20}{3} = \dfrac{180}{\square}$ c $\dfrac{11}{2} = \dfrac{99}{\square}$

d $\dfrac{32}{7} = \dfrac{\square}{35}$ e $\dfrac{100}{28} = \dfrac{\square}{7}$ f $\dfrac{144}{18} = \dfrac{24}{\square}$

2 a How many fifths are in $3\tfrac{4}{5}$?   b How many thirds are in $7\tfrac{1}{3}$?
   c How many twelfths are in $4\tfrac{7}{12}$?   d How many sevenths are in 32?

3 Write each of the following as a mixed number in its simplest form.

a $\tfrac{13}{5}$ b $\tfrac{32}{3}$ c $\tfrac{28}{9}$ d $\tfrac{40}{6}$ e $\tfrac{60}{8}$
f $\tfrac{84}{18}$ g $\tfrac{133}{21}$ h Thirteen thirds   i Eighteen eighths

4 Write each of the following as a fraction or mixed number.

a The fraction of a litre given by each of the following.
   i 320 cl   ii 645 cl   iii 272 cl   iv 9375 ml
b The fraction of a foot (ft) given by each of the following.
   i 20 in   ii 51 in   iii 100 in   iv 114 in

Practice

12B Adding and subtracting fractions

If necessary, convert your answers to mixed numbers and cancel down.

1 Use an eighths fraction chart to calculate the following.

a $\tfrac{7}{8} + \tfrac{3}{4}$ b $1\tfrac{5}{8} + 2\tfrac{7}{8}$ c $1\tfrac{1}{4} - \tfrac{3}{8}$ d $3\tfrac{1}{2} - 1\tfrac{5}{8}$ e $2\tfrac{1}{4} + 1\tfrac{3}{8} - 2\tfrac{5}{8}$

2 Calculate the following.

a $\tfrac{2}{7} + \tfrac{3}{7}$ b $\tfrac{11}{12} - \tfrac{7}{12}$ c $\tfrac{4}{5} + \tfrac{3}{5} + \tfrac{4}{5}$ d $\tfrac{5}{8} + \tfrac{7}{8} - \tfrac{3}{8}$

3 Convert the fractions to equivalent fractions with a common denominator. Then calculate the answer.

a $\tfrac{1}{5} + \tfrac{1}{2}$ b $\tfrac{1}{10} + \tfrac{2}{5}$ c $\tfrac{5}{6} + \tfrac{4}{9}$ d $\tfrac{3}{8} + \tfrac{5}{6} + \tfrac{1}{4}$
e $\tfrac{3}{4} - \tfrac{1}{3}$ f $\tfrac{11}{12} - \tfrac{3}{4}$ g $\tfrac{7}{9} - \tfrac{1}{3}$ h $\tfrac{9}{10} - \tfrac{1}{2} - \tfrac{1}{5}$

4 $\frac{3}{5}$ of Jan's emails are junk mail and $\frac{1}{10}$ is from friends. The rest is work related.

    a What fraction is work related?
    b Jan received 120 emails during the week. How many were not junk mail?

5 A poster is printed using red, blue and yellow inks. Of the ink used, $\frac{2}{9}$ is red and $\frac{1}{6}$ is blue.

    a What fraction of the ink used is not yellow?
    b 36 ml of ink is used to print the poster. Calculate the amount of each ink used.

Practice
12C Order of operations

Do not use a calculator. Show all of your working.

1 Write down the operation that you would do first in each calculation. Then calculate the answer.

    a $12 - 3 \times 2$ **b** $2 \times (9 - 5)$ **c** $10 \times 2 \div 5 + 3$
    d $30 - 20 + 10$ **e** $12 + 8 - 3^2$ **f** $4 \times (2 + 5)^2$

2 Calculate the following. Show each step of your calculation.

    a $3^2 + 5 \times 2$ **b** $10 - (1 + 2)^2$ **c** $3 \times 12 \div 3^2$
    d $32 \div (3^2 - 1)$ **e** $\dfrac{60 + 12}{2 \times 3}$ **f** $\dfrac{60}{(6 + 3^2)}$
    g $1.5 + 3 \times (2.4 - 0.8) - 2.1$

3 Copy each calculation. Insert brackets to make the answer true.

    a $11 - 7 - 1 + 4 = 1$ **b** $1 + 4 + 3^2 = 50$
    c $24 \div 2 \times 3 = 4$ **d** $6 + 9 \div 12 \div 4 = 5$
    e $12 - 3^2 - 7 \times 4 = 4$

4 Calculate the following. Work out the inside (round) brackets first.

    a $120 - [70 - (90 - 55)]$ **b** $[(2 + 7)^2 - 12] \div 3$
    c $(2 + 6) \times [36 \div (9 - 5)]$ **d** $12^2 - [5^2 - 2 \times (32 - 27)]$

Practice
12D Multiplying decimals

Do not use a calculator.

1 Calculate the following.

    a 0.6×0.7 **b** 0.9×0.9 **c** 0.05×0.7
    d 0.8×0.03 **e** 0.4^2 **f** 0.04×0.09
    g 0.003×0.6 **h** 0.008×0.002

2 Calculate the following.

    a 200×0.9 **b** 0.8×400 **c** 500×0.09
    d 2000×0.7 **e** 0.006×600 **f** 70×0.04
    g $90\,000 \times 0.004$ **h** 3000×0.002

3 A seed weighs 0.04 g. How much does a box of 600 seeds weigh if the box alone weighs 17 g?

4 Calculate the following.

    a $300 \times 0.4 \times 0.8$ **b** $0.006 \times 7000 \times 0.2$ **c** $0.04 \times 0.07 \times 300$

5 Sound travels about 0.3 km in 1 second. How far does sound travel in the following times?

    a 200 seconds   **b** 10 minutes   **c** 0.02 seconds

Work in kilometres.

Practice

12E Dividing decimals

Do not use a calculator.

1 Calculate the following.

    a $0.6 \div 0.03$ **b** $0.08 \div 0.2$ **c** $0.36 \div 0.04$
    d $0.9 \div 0.4$ **e** $0.04 \div 0.001$ **f** $0.15 \div 0.003$
    g $0.07 \div 0.05$ **h** $0.8 \div 0.002$

2 Calculate the following.

    a $9 \div 0.3$ **b** $40 \div 0.8$ **c** $60 \div 0.03$
    d $48 \div 0.06$ **e** $500 \div 0.02$ **f** $900 \div 0.003$
    g $5000 \div 0.5$ **h** $120 \div 0.06$

3 Calculate the following.

    a $2.8 \div 20$ **b** $32 \div 800$ **c** $1.2 \div 300$
    d $0.16 \div 400$ **e** $0.54 \div 90$ **f** $0.008 \div 400$

4 0.008 g of platinum costs £0.16. Calculate the cost of 1 g. Work in £.

5 An advertiser pays 0.004p to a website every time it is hit (a hit is when someone visits the website). In one month, the advertiser pays the website £16. How many hits did the website receive?

6   **a** 3000 bacteria have a mass of 0.000 09 g. What is the mass of one bacterium? (Bacteria is the plural of bacterium.)

       b 10 000 atoms have the same mass as one bacterium. What is the mass of an atom?

CHAPTER 13 Algebra 5

Practice 13A Expand and simplify

Simplify the following expressions.

1
- **a** $4p + 7p$
- **b** $9u - 3u$
- **c** $7d - d$
- **d** $3i + 5i + 8i$
- **e** $4n + 5n - 5n$
- **f** $10h - 7h - h$
- **g** $2f - 7f$
- **h** $-6y - 3y$
- **i** $2d - 5d - 5d$

2
- **a** $7s + 2s + 4t$
- **b** $9i - 4i + 2j$
- **c** $4a + 3b + 2b$
- **d** $5d + 4y + 3d$
- **e** $6r + 3h - 4r$
- **f** $5b + 3d - 7d$
- **g** $5u + 2a + 3u + 4a$
- **h** $4d + 5y - 2y + 3d$
- **i** $10k + 4p - 7k - 2p$
- **j** $6t - 3g + 2t - 4g$

3 Expand the brackets.
- **a** $4(x + 8)$
- **b** $7(3d - 5)$
- **c** $2(4f + 2e)$
- **d** $d(3 - u)$
- **e** $m(2r + c)$
- **f** $j(3h - 2g)$

Expand the brackets and then simplify the following expressions.

4
- **a** $2x + 3(x + 5)$
- **b** $4d + 2(5d + 6)$
- **c** $9i - 4(i + 2)$
- **d** $7u - 3(2u - 3)$
- **e** $6k + 2e + 3(2k + e)$
- **f** $4a + 7b - 2(5a - 2b)$

5
- **a** $3(4i + 2) + 4(i + 2)$
- **b** $3(2s + 3) + 2(s - 4)$
- **c** $2(3i - 5) + 3(4i - 1)$
- **d** $4(2s + 3) - 3(s + 2)$
- **e** $5(2f + 2) - 3(3f - 3)$
- **f** $4(3 + 2g) - 3(2 - 4g)$

Practice 13B Solving equations

Solve the following equations

1
- **a** $3t = 18$
- **b** $4u = 32$
- **c** $9p = 27$
- **d** $2h = 13$
- **e** $10y = 76$
- **f** $5m = -30$
- **g** $-2x = 10$
- **h** $-5m = -20$

2
- **a** $4 = \dfrac{8}{p}$
- **b** $5 = \dfrac{20}{b}$
- **c** $7 = \dfrac{35}{f}$
- **d** $10 = \dfrac{70}{e}$
- **e** $2 = \dfrac{7}{y}$
- **f** $5 = \dfrac{23}{w}$

3
- **a** $2x + 1.5 = 11.5$
- **b** $4.5y - 2.5 = 6.5$
- **c** $1.2i - 1.2 = 2.4$
- **d** $1.5n + 8 = 2$

Solve the following equations. Expand the brackets first.

4
a $4(d + 3) = 32$
b $3(2f - 1) = 9$
c $5(2s + 1) = 30$
d $3(2g + 10) = 6$
e $4(10 - t) = 12$
f $3(5 - 4q) = 39$

5
a $2(d + 1) + 3(d + 2) = 23$
b $3(2s + 5) + 4(s - 3) = 43$
c $3(3w + 2) - 2(2w - 1) = 28$
d $2(2k + 3) - 3(k + 1) = 1$

Practice — 13C Constructing equations to solve

1
a A book and writing pad cost £8 altogether. The book costs £x. Write an expression for the cost of the writing pad.
b Maria is twice the age of Janine. Janine is j years old. Write an expression for the age of Maria.
c The difference between the heights of two trees is 4 m. The taller tree is T metres high. Write an expression for the height of the shorter tree.

2 Solve each of the following problems by creating an equation and then solving it.

a Tyres cost £x each. Four tyres cost £112. Find the cost of one tyre.
b Jamie is y years old and Paul is 9 years old. Their ages total 16 years. How old is Jamie?
c The difference between two numbers is five. The smaller number is n and the larger number is 12. What is the smaller number?
d A plum weighs p grams. A lemon weighs 12 g more than the plum.
  i Write an expression for the total weight of the lemon and plum.
  ii If the total weight is 44 g, find the weight of the plum.
e A soap opera lasts three times as long as a cartoon. The cartoon lasts m minutes.
  i Write an expression for the total length of the programmes.
  ii The soap opera and the cartoon last 32 minutes altogether. How long does the cartoon last?
f The sum of three consecutive numbers is 78. The smallest number is n. Find the value of n.
g I am x years old. Three times my age four years ago is 27. How old am I?
h A bakery sold 15 boxes of cakes. Large boxes contain four cakes and small boxes contain three cakes. 51 cakes were sold altogether. How many large boxes were sold? (Hint: Let the number of large boxes be x.)

Practice — 13D Problems with graphs

1 Find the gradient of each of the following lines.

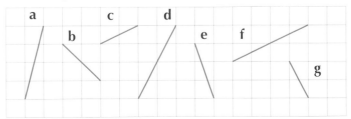

2 Find the gradient of the straight line that joins each pair of coordinates. Plot the points using squared paper.

    a (1, 4) and (3, 10)   **b** (2, 10) and (4, 2)   **c** (0, 3) and (6, 6)
    d (−3, 2) and (0, 8)   **e** (−1, 6) and (1, −6)

3 Find the gradient and *y*-axis intercept for each of the following equations.

    a $y = 3x + 5$ **b** $y = -2x + 3$ **c** $y = 0.5x - 3$ **d** $y = 3x$

4 Write the equation of each line in the form $y = mx + c$, given the following information.

    a $m = 5$, $c = 2$ **b** Gradient is 6, *y*-intercept is −3
    c Gradient is −3, *y*-intercept is 9   **d** Gradient is 1.2, *y*-intercept is 0

5 Find the gradient, *y*-intercept and equation for each graph below.

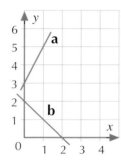
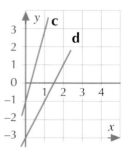
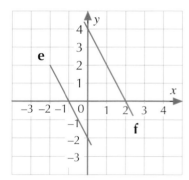
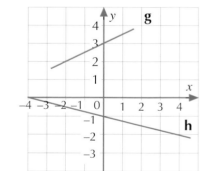

13E Real-life graphs

1 Sketch a graph to illustrate each of the following situations. Estimate any necessary measurements.

    a The height of a person from birth to age 30 years.
    b The temperature in summer from midnight to midnight the next day.
    c The height of water in a WC cistern from before it is flushed to afterwards.

2 The graph below shows the cost of car hire for two companies.

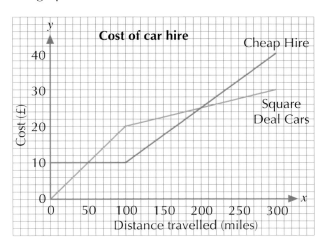

a Estimate the cost of hiring a car from each company and travelling the following distances.
  i 60 miles   **ii** 100 miles   **iii** 240 miles
b Describe the way each company charges for car hire.
c When is Cheap Hire cheaper than Square Deal Cars?

3 Match each description to its graph.

  a The temperature of the desert over a 24-hour period.
  b The temperature of a kitchen over a 24-hour period.
  c The temperature of a pond during a month of winter.
  d The temperature of a cup of tea as it cools down.

i
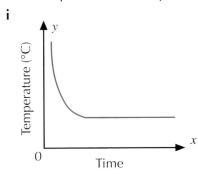

ii
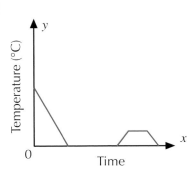

iii
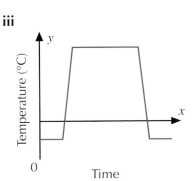

iv
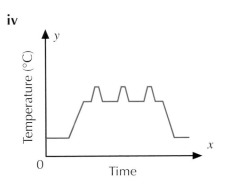

FM **4** George tried several diets to lose weight. The table shows his weight loss over a one-year period. He weighed 105 kg at the beginning of the year.

Diet	Weight loss/gain
Hay diet	Lost 10 kg in 2 months
Calorie counter	Stayed same weight for 3 months
Vegetarian	Gained 3 kg in 2 months
Usual diet	Gained 12 kg in 1 month
Atkins diet	Lost 28 kg in 4 months

Draw a graph showing how George's weight changed over time. Start the vertical weight axis at 80 kg and use a scale of 1 cm to 2 kg.

Practice 13F Change of subject

1 Rewrite each of the following formulae as indicated.

   a $T = D + 4$ Express D in terms of T.
   b $m = \frac{d}{5}$ Express d in terms of m.
   c $P = 6T$ Express T in terms of P.
   d $R = 7 - s$ Express s in terms of R.

2 Rewrite each of the following formulae as indicated.

   a $A = 9p$ Make p the subject of the formula.
   b $y = x + 5$ Make x the subject of the formula.
   c $A + 5 = C$ Make A the subject of the formula.
   d $y = 8 + c$ Make c the subject of the formula.
   e $T = 4m$ Make m the subject of the formula.

3 The perimeter of a shape is given by the formula $P = a + 5$.

   a Find the value of P when $a = 4$ cm.
   b Make a the subject of the formula.
   c Calculate the value of a when $P = 37$ cm.

4 The speed of a car is u mph. It accelerates to a speed of v mph. Its final speed is given by the formula $v = u + 16$.

   a Calculate the final speed of the car if it accelerates from 20 mph.
   b Make u the subject of the formula.

5 $A = a + 4$

   a Find A when $a = 5$.
   b Make a the subject of the formula.
   c $A = 18$. Use your formula to find a.

CHAPTER 14 Solving Problems

Practice

14A Number and measures

1. Two pens and a pencil have a total length of 49 cm. Four pens and a pencil have a total length of 77 cm. What is the total length of seven pens and a pencil?

2. Complete the puzzle below so that each large triangle adds up to 50.

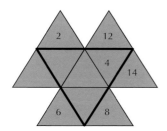

3. Harry's car can travel 320 miles on a full tank of petrol. He drives 124 miles from Harwich to Dover, where he catches a ferry, then 330 km from Calais to the Hook of Holland. Does he need to stop for petrol?

4. a Find three consecutive even numbers that add up to 114.
   b Find two consecutive even numbers with a product of 1088.

5. A bag of potatoes in the UK weighs 15 lb. A bag of potatoes in France weighs 7 kg. Which is heavier?

6. Two 12-hour electric clocks start at 12 o'clock. Clock A loses 5 minutes every hour. Clock B keeps perfect time.
   a What will be the time showing on clock A when clock B next reaches 12 o'clock?
   b How long after the start time will it be before both clocks show 12 o'clock again?

Practice
14B Using algebra, graphs and diagrams to solve problems

1 A pond is 240 cm deep. The depth of the pond decreases by 2 cm every day.
  a How long before the pond is half its original depth?
  b Write down a formula to show the depth D of the pond in n days.

2 I think of a number, treble it and then subtract 24. The answer is 27. What was the number I first thought of?

3 A fruit bowl weighs 500 g and holds oranges that weigh 200 g each.
  a Write a formula to show the total weight W of a bowl containing n oranges.
  b Draw a graph to show the total weight of the fruit bowl and oranges. Use the x-axis for the number of pieces of fruit, numbered from 0 to 7. Use the y-axis for total weight, numbered from 0 to 2000 g.

Another fruit bowl weighs 800 g and holds apples that weigh 100 g each.

  c *Using the same axes*, draw another graph to show the total weight of a bowl of n apples.
  d Use your graphs to find when a bowl of oranges weighs the same as a bowl of apples.

4 Five times the number I am thinking of is 16 more than treble the number. What is the number I am thinking of?

5 a Fold an A4 sheet of paper into three equal parts (see diagram **i**). Fold the folded sheet in half (see diagram **ii**). Open the sheet and count the number of small rectangles.

i 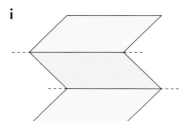   **ii**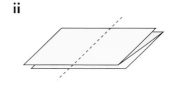

  b Repeat part **i**, and then fold the folded sheet in half *twice*. Open the sheet and count the rectangles.
  c Repeat part **i**, and then fold the folded sheet in half *three times*. Open the sheet and count the rectangles.
  d Write down a formula for the number N of rectangles made by folding the folded sheet (**i**) in half n times.
  e How many rectangles would be made by folding the folded sheet (**i**) in half five times?
  f How many times would you have to fold the folded sheet (**i**) in half to make 200 rectangles or more?

Practice 14C Logic and proof

1 Copy and complete the following number problems.

a ☐ 5 ☐
 + 1 ☐ 7
 ─────
   4 4 6

b   3 ☐
  × ☐ 4
  ─────
   8 8 8

c $7\square^2 = \square 9 \square 9$

d $(10 + \square)(\square + 8) = 182$

2 It takes 8 people 12 hours to dig a hole. How long would it take the following numbers of people?

 a 4 people   b 2 people   c 6 people

3 Six glasses hold 120 cl altogether. Eight mugs hold 144 cl altogether. Which holds more: a mug or a glass?

4 a Give an example to show that the product of three odd numbers is odd.
 b Prove that the product of three odd numbers is odd.
  (Hint: Remember the product of two odd numbers.)

5 a Give an example to show that the product of two odd and two even numbers is even.
 b Prove that the product of two odd and two even numbers is even.

6 a Write down the factors of 21.
 b Prove that all of the factors of an odd number are odd.

Practice 14D Proportion

1 Two of the carriages on a train are first class. The other six carriages are second class.

 a What proportion of carriages are first class?
 b What is the ratio of first-class to second-class carriages?
 c A train with 12 carriages has the same proportion of first-class carriages. How many first-class carriages does it have?

2 The ratio of nitrogen to oxygen in the air is approximately 4 : 1. A cupboard contains 600 litres of air.

 a How much nitrogen does it contain?
 b How much oxygen does it contain?

3 Eight light bulbs cost £7.20. How much does a box of 20 cost?

4) There are three females for every four males in a social club with 91 members. How many females are there?

5) Two gallons is approximately 9 litres.
   a How many gallons are equivalent to 63 litres?
   b How many litres are equivalent to 7.5 gallons?

6) Eight grams of silver are used to make 15 cm of chain.
   a How much silver does 21 cm of chain contain?
   b How long is a chain that contains 25 grams of silver?

7) A photo of area 108 cm^2 is enlarged to have an area of 180 cm^2. What is the ratio of the two areas?

Practice
14E Ratio

1) Simplify the following ratios.
   a 12 : 9   b 16 : 30   c 18 kg : 42 kg
   d £4 : 75p   e 8 weeks : 12 days f 45 cm to 3 m
   g 5 kg : 1.75 kg

2)
   a Divide 40 cm in the ratio 7 : 1.
   b Divide 600 mm in the ratio 11 : 9.
   c Divide 5000 people in the ratio 3 : 5.
   d Divide £76 in the ratio 2 : 7 : 10.

3) Donna eats three times more hot meals than cold meals. She eats 96 meals in July. How many of them were hot?

4) The ratio of children in Year 8 who own a mobile phone to those that don't is 3 : 2. There are 210 children in Year 8. How many children do not own a mobile phone?

5) The land area to sea area of the surface of the Earth is in the ratio 3 : 7. The total surface area of the Earth is 500 000 000 km^2. What is the area of land?

CHAPTER 15 Geometry and Measures 4

Practice

15A Plans and elevations

1 Draw each of the following 3-D shapes on an isometric grid.

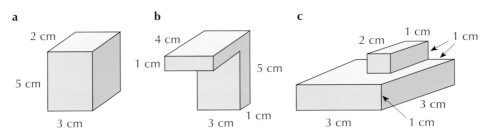

2 For each of the following 3-D shapes, draw a:

    i plan   **ii** front elevation   **iii** side elevation.

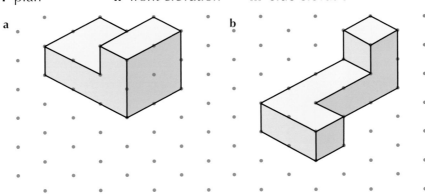

3 The plan, front and side elevations of a 3-D shape are shown below. Draw the solid on an isometric grid.

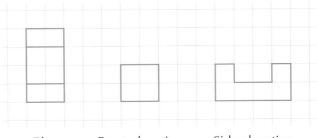

Practice 15B Scale drawings

1 These objects have been drawn using the scales shown. Find the true lengths of the objects. In **a**, measure the length of the golf club shaft only.

a

1 cm to 10 cm

b

2 cm to 1 m

c

1 cm to 0.7 m

d

2 cm to 3 m

2

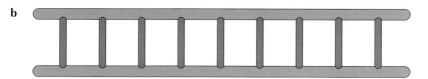

Scale: 1 cm to 120 m

The diagram shows a scale drawing of an aircraft hanger (plan view).

a Calculate the real length of the hanger.
b Calculate the real width of the hanger.
c Calculate the area of the hanger.

3 Copy and complete the table.

	Scale	Scaled length	Actual length
a	1 cm to 2 m		24 m
b	1 cm to 5 km	9.2 cm	
c		6 cm	42 miles
d	5 cm to 8 m	30 cm	

Scale: 2 cm to 25 m

			Deliveries
Wine	Delicatessen	Bakery	Frozen food
Fresh food	Packaged food	Non-food items	Chemist
	Checkout area		

The diagram shows a scale drawing of a supermarket. Make a table showing the dimensions and area of each section of the supermarket. (Hint: With the two L-shaped areas, divide them into two rectangles.)

Practice

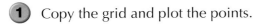
15C Finding the mid-point of a line segment

1 Copy the grid and plot the points.

a Make a list of the coordinates of the points.

b Join these pairs of points using straight lines.
  - i   AB
  - ii AD
  - iii DE
  - iv EF
  - v   AG
  - vi EH
  - vii BE

c Find the mid-point of each line in part **b**.

d Make a list of the coordinates of the mid-points.

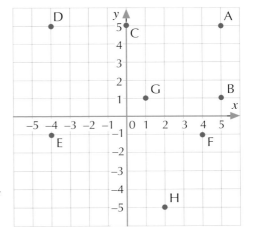

2 Copy the grid and plot the points.

a B is the mid-point of line AC. Find the coordinates of point C.

b B is the mid-point of line DE. Find the coordinates of point E.

c F is the mid-point of line EG. Find the coordinates of point G.

d J is the mid-point of line DK. Find the coordinates of point K.

e The origin is the mid-point of KL. Find the coordinates of point L.

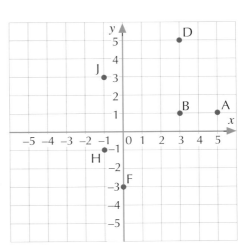

3 Find the mid-points of the following line segments. Use any method you know (diagram or calculation).

   a A(1, 3) joined to B(5, 7)
   b C(0, 4) joined to D(5, 8)
   c E(–4, 0) joined to F(0, 6)
   d G(–6, –2) joined to H(2, –4)

Practice
15D Constructing triangles

1 Construct the following triangles. Label your triangles.

   a   **b**   **c**

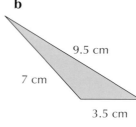
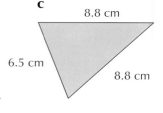

2 Construct each of the following triangles. Describe the type of triangle you have drawn.

   a △ABC where AB = 9 cm, BC = 6 cm and AC = 6 cm
   b △PQR where PQ = 7.2 cm, QR = 3.4 cm and PR = 9.1 cm

3 Construct the following triangles using ruler, protractor and compasses.

   a   **b**   **c**

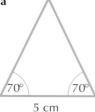
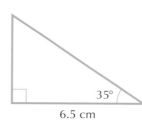
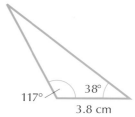

Practice
15E Circumference and area of a circle

In this exercise, take π = 3.14 or use SHIFT x10ˣ on your calculator.

1 Calculate the circumference of each of the following circles. Give each answer to one decimal place.

   a   **b**   **c**   **d**

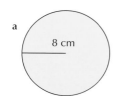
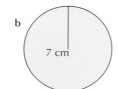
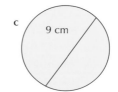
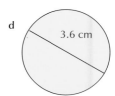

2 Calculate the area of each of the following circles. Give each answer to one decimal place.

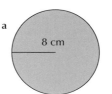
a

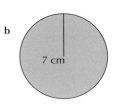
b

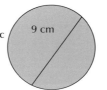
c

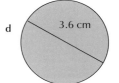
d

3 A CD has a diameter of 12 cm. Calculate its circumference and area. Give your answers to one decimal place.

Practice 15F Bearings

1 Use a protractor to find the three-figure bearings of B from A.

a   b   c

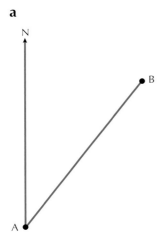

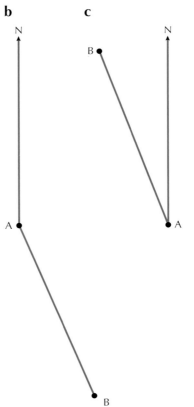

2 Find the three-figure bearings of P from Q.

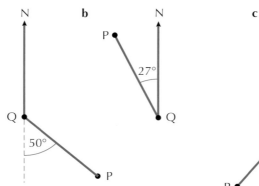

3 Sketch the following bearings. Label your diagrams.

    a A skier is on a bearing of 72° from the ski lodge.
    b Clerkhill is 12 miles from Moffat on a bearing of 115°.

4

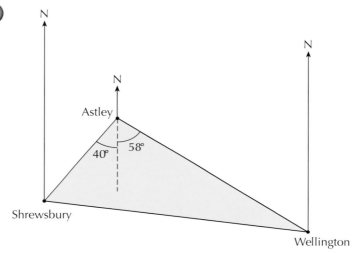

Calculate the three-figure bearing of each of the following.

    a Wellington from Astley   **b** Astley from Wellington
    c Astley from Shrewsbury   **d** Shrewsbury from Astley

Use your protractor to find the three-figure bearing of the following.

    e Wellington from Shrewsbury   **f** Shrewsbury from Wellington

CHAPTER 16 Statistics 3

Practice
16A Frequency tables

1 The table shows the lengths (L metres) of 30 pythons.

Length of snake (L metres)	Frequency
$3.5 \leq L < 4$	2
$4 \leq L < 4.5$	4
$4.5 \leq L < 5$	7
$5 \leq L < 5.5$	9
$5.5 \leq L < 6$	5
$6 \leq L < 6.5$	3

a One of the snakes is 4.5 metres long. Which class contains this length?
b How many snakes are shorter than 5 metres?
c How many snakes have a length of 5.5 metres or more?
d How many snakes are between 4 and 6 metres long?
e How many snakes could be exactly 5 metres long?

2 The volumes (V cl) of liquid contained in 20 coconuts are shown below.

12.2	11.1	10.5	12.8	12.0	10.1	11.8	12.3	10.7	12.7
10.0	11.6	12.1	10.5	10.8	12.6	10.7	11.4	12.8	11.3

Copy and complete the table.

Volume of liquid (V cl)	Tally	Number of coconuts
$10 \leq V < 10.5$		
$10.5 \leq V < 11$		
$11 \leq V < 11.5$		
$11.5 \leq V < 12$		
$12 \leq V < 12.5$		
$12.5 \leq V < 13$		

3 The durations (t minutes) of 30 telephone calls are shown below.
Times have been rounded up to the nearest minute.

4	12	8	1	19	7	7	28	14	54
9	2	20	16	2	43	5	18	1	5
5	14	9	10	3	30	11	6	17	2

Copy and complete the following frequency table.

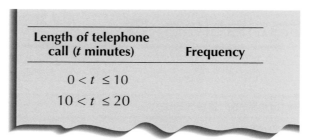

Length of telephone call (*t* minutes)	Frequency
$0 < t \leq 10$	
$10 < t \leq 20$	

Practice

16B Assumed mean and working with statistics

1 a Find the mean of 256, 251, 259, 250, 255, 255, 253, 249.
  Use 250 as assumed mean.
 b Find the mean of 72.7, 72, 72.9, 71.7, 72.8, 72.3, 72.6, 72.2, 72.6.
  Use 72 as assumed mean.
 c Find the mean of 58 284, 58 280, 58 287, 58 278, 58 289.
  Use 58 280 as assumed mean.

2 Write down two numbers with a range of 4 and a mean of 7.

3 Write down three numbers with a mode of 10 and a mean of 9.

4 The mean of four numbers is 5, the mode is 1 and the range is 11. What are the four numbers?

5 The mode of a set of numbers is 5, their range is 10 and their mean is 7. Each of the numbers is increased by 10.

  a How are the averages affected?
  b What is the new range?

6 The weekly wages of four employees are £320, £290, £420 and £370.

  a Calculate the mean and range for the wages.
  b The employer hires a fifth worker. She wants the mean wage to be £340. What should she pay the fifth worker?
  c All workers receive a pay rise of £30 per week. How will this affect the mean and range? (Do not recalculate them.)
  d The following year the workers receive a 10% pay rise. How do you think the mean and range will be affected?

Practice 16C Drawing frequency diagrams

1 Draw a bar chart for each of the following frequency tables.

a Amounts of beer served in 500 ml glasses are shown in the table below.

Amount of beer (V ml)	Number of glasses
$485 \leq V < 490$	3
$490 \leq V < 495$	7
$495 \leq V < 500$	13
$500 \leq V < 505$	18
$505 \leq V < 510$	11

b Shop prices for the same beach ball are as follows.

Price of beach ball (£P)	Number of shops
£2.80 < P ≤ £3	27
£3 < P ≤ £3.20	51
£3.20 < P ≤ £3.40	30
£3.40 < P ≤ £3.60	23
£3.60 < P ≤ £3.80	13
£3.80 < P ≤ £4	9

2 The table shows the average height of sweetcorn plants after being sprayed with different amounts of a new fertilizer.

Amount of fertilizer (A ml)	Height of plant (h m)
0	1.40
10	1.40
20	1.45
30	1.55
40	1.70
50	1.70
60	1.60
70	1.50
80	1.40
90	1.30
100	1.30
110	1.30

a Plot a graph. Use the following scales:
   x-axis (Amount of fertilizer): 1 cm to 10 ml
   y-axis (Height of plant): 2 cm to 0.10 m
b Estimate the level of fertilizer that would give plants a height of 1.50 m.
c Estimate the height of a plant sprayed with 35 ml of fertilizer.
d Which level of fertilizer would you advise the farmer to use? Explain your answer.
e At which levels did the fertilizer not improve growth?

Practice

16D Comparing data

1 QuickDrive and Ground Works are two companies that lay drives. The numbers of days each company takes to complete nine drives are shown below.

QuickDrive 2, 5, 3, 3, 6, 2, 1, 8, 3
Ground Works 3, 2, 3, 1, 4, 3, 2, 2, 1

a Calculate the mode, mean and range for each company.
b Comment on the differences between the averages.
c Comment on the difference between the ranges.

2 The table shows the weights of fish (in grams) caught by three anglers in a competition.

Jerry	230	100	380	720	450			
Aditya	230	400	280	320	250			
Marion	100	130	430	200	70	180	200	90

a Calculate the mean and range for each of the anglers.
b Who was the most consistent? Explain your answer.
c Who performed the best overall? Explain your answer.
d Which angler would you choose to enter a competition that offered prizes for the heaviest fish caught? Explain your answer.

3 The table shows the mean and range for the average weekly rainfall (mm) in two holiday resorts.

	Larmidor	Tutu Island
Mean	6.5	5
Range	33	62

Explain the advantages of each island's climate using the mean and range.

Practice

16E Which average to use?

1 For each set of data, do the following.

i Calculate the given average.
ii If the chosen average is unsuitable, give a reason why.

a Mode   5, 5, 7, 9, 9, 9, 13
b Mean   23, 25, 26, 29, 31, 36, 40
c Median 11, 13, 15, 15, 16, 16, 16, 18, 18, 20, 80
d Mean   0.3, 0.3, 2.3, 2.6, 2.9, 3.0
e Median 120, 125, 140, 170, 1200, 1300, 2000

2 For each set of data, decide whether the range is a suitable measure of spread or not.

    a 30, 60, 100, 120, 150, 200   **b** 13, 45, 48, 52, 66
    c 5, 5, 5, 15, 25, 25, 25, 25, 25

3 Judges awarded the following points to competitors in a surfing competition.

    23, 6, 47, 29, 41, 17, 38, 21, 30, 24, 40, 7, 27, 18, 20, 6, 26, 35, 28

    a Calculate the mean, median and mode.
    b Which measure(s) of average are suitable?
    c Which measure(s) of average are unsuitable? Explain your answer.
    d Copy and complete the tally chart.

Points awarded, p	Tally	Frequency
$0 \leq p < 10$		
$10 \leq p < 20$		

    e   **i** State the modal class.
         ii Is this a suitable average?

4 Two shops offer the following sizes of dresses.

Periwinkle	12, 14, 16, 18, 30
Jenny's	10, 12, 14, 16, 18, 20, 22

    a Which shop has the greater range of sizes?
    b Do you think this is a suitable measure? Explain your answer.

Practice
16F Experimental and theoretical probability

1 People can choose a raffle ticket number from those below.

    1   2   3   4   5   6   7   8   9   10
    11   12   13   14   15   16   17   18   19   20
    21   22   23   24   25   26   27   28   29   30

    a What is the theoretical probability of someone choosing a number from the middle row?
    b Design and carry out an investigation to test the following hypothesis:
        People are most likely to choose a number from the middle row.

2 **a** Roll a die until you get an odd number. What is the theoretical probability that the next roll will be odd too?
    b Design and carry out an experiment to test the following hypothesis:
        An odd number is less likely to be followed by another odd number.